The Effects of Bilingualism on Non-Linguistic Cognition

Jennifer Mattschey

The Effects of Bilingualism on Non-Linguistic Cognition

A Historic Perspective

Jennifer Mattschey
Faculty of Arts and Social Sciences
The Open University
Milton Keynes, UK

ISBN 978-3-031-34680-4   ISBN 978-3-031-34681-1   (eBook)
https://doi.org/10.1007/978-3-031-34681-1

© The Editor(s) (if applicable) and The Author(s), under exclusive licence to Springer Nature Switzerland AG 2023
This work is subject to copyright. All rights are solely and exclusively licensed by the Publisher, whether the whole or part of the material is concerned, specifically the rights of translation, reprinting, reuse of illustrations, recitation, broadcasting, reproduction on microfilms or in any other physical way, and transmission or information storage and retrieval, electronic adaptation, computer software, or by similar or dissimilar methodology now known or hereafter developed.
The use of general descriptive names, registered names, trademarks, service marks, etc. in this publication does not imply, even in the absence of a specific statement, that such names are exempt from the relevant protective laws and regulations and therefore free for general use.
The publisher, the authors, and the editors are safe to assume that the advice and information in this book are believed to be true and accurate at the date of publication. Neither the publisher nor the authors or the editors give a warranty, expressed or implied, with respect to the material contained herein or for any errors or omissions that may have been made. The publisher remains neutral with regard to jurisdictional claims in published maps and institutional affiliations.

Cover illustration: Pattern © Harvey Loake

This Palgrave Macmillan imprint is published by the registered company Springer Nature Switzerland AG.
The registered company address is: Gewerbestrasse 11, 6330 Cham, Switzerland

Preface

Historically, bilingualism has been linked to such undesirable terms as "mental retardation" and "mental confusion". Much of this early research used unreliable methods to test if bilingualism affects intelligence, at times influenced by eugenics theories. This early research coined the term "Bilingual Problem", yet the supposed problem of bilingualism was later found to be the result of poor control for confounding variables (e.g. socio-economic background) and inappropriate choices of intelligence tests to compare bilinguals and monolinguals. These days, we know that the ability to speak more than one language has no negative effects on intelligence—in fact, it appears to enhance executive functioning. But does it really? Recent research suggests that findings that point towards an executive functioning advantage for bilinguals have been as much affected by poor control of confounding variables and sub-optimal use of available tests as early research on intelligence. In fact, there is strong evidence to suggest that there is no difference between bilinguals and monolinguals after all. Are we reluctant to let go of the idea of a bilingual advantage because of the history of research on the effects of bilingualism on non-linguistic cognition?

Milton Keynes, UK   Jennifer Mattschey

Acknowledgements

This book would not exist if my husband, tired of my complaints that such a book does not already exist, had not encouraged me to go ahead and submit the proposal for it to Palgrave. I have to thank both him and my editor, Beth Farrow, for giving me the opportunity to write it. Additionally, I would like to thank my reviewers for their insightful and constructive comments.

Contents

1 **Introduction**   1
   1.1 *Suppressed Languages and Education*   3
   1.2 *Who Is Bilingual and Who Is Monolingual?*   5
   1.3 *Unimodal and Bimodal Bilingualism*   13
   1.4 *The Bilingual Mind*   15
   1.5 *Summary of Bilingualism*   20
   References   20

2 **Bilingual Education in the Early Twentieth Century**   27
   2.1 *Early Bilingual Education Reports*   27
   2.2 *Early Case Studies of Bilingual Children*   33
   References   34

3 **The Bilingual Problem**   37
   3.1 *The Bilingual Problem*   37
   3.2 *Early Intelligence Tests*   40
   3.3 *Late 1920s and 1930s*   47
   References   51

4 **Mid-Twentieth Century: Bilingualism and Intelligence**   55
   4.1 *1940s*   55
   4.2 *1950s*   56
   4.3 *1960s*   59
   References   66

5	**Late Twentieth Century: Meta-Linguistics**	69
	5.1 1970s	69
	5.2 1980s	71
	5.3 1990s	74
	References	77
6	**The Bilingual Advantage**	81
	6.1 *The Bilingual Advantage*	81
	6.2 *Confirmatory Bias and Selective Reporting*	84
	6.3 *Bilingualism and Dementia*	92
	6.4 *To Match Groups or Not Match Groups*	94
	6.4.1 *Immigration*	94
	6.4.2 *Socio-economic Status*	98
	6.4.3 *Types of Bilingualism*	104
	6.5 *Further Considerations: Could We Control for All Potentially Confounding Variables?*	106
	References	108
7	**Is Bilingualism Good or Bad?**	119
	7.1 *The Same Old Question?*	120
	7.2 *Theoretical Frameworks*	122
	7.3 *Looking Ahead*	123
	References	125
	Index	129

CHAPTER 1

Introduction

Abstract The language we choose to speak can give someone else the opportunity to join in the conversation if it is a language we share with them. In contrast, language can also exclude someone from a conversation if they do not speak the language the conversation is held in. For bilinguals, this often means they switch between languages as needed. This chapter provides an overview of different types of bilingualism and discusses evidence that shows languages are active in parallel in bilinguals.

Keywords Bilingualism • Proficiency • Language use • Language switching

If we listen to a conversation in a foreign language, we may be able to draw certain conclusions. The sound of the voices we hear may suggest they are angry, or happy, or sad. The speaker's body language and facial expressions may be supportive and loving, or aggressive and cold. While listening in, we may notice if they take turns to speak (Who speaks more? Who is the better listener?) or talk over each other. The voices may sound young or old, female or male, sober or slurred, rough or gentle. Just by listening to or watching a conversation, we can derive a fair amount of information about the speakers and possibly about the content of that conversation. However, if we do not understand the words being spoken, that information remains limited. As such, people in multilingual societies will frequently find themselves switching between languages to adapt to the people present around them. Language can invite us into a conversation as quickly as it can exclude us from it.

Of course, not everybody speaks more than one language or aspires to be able to do so. Yet, being bilingual is often not something we actively choose to become but is rather the result of our circumstances. A newborn does not choose to be born into a multilingual society or to parents who speak different native languages, yet they might acquire the languages they are surrounded by just as naturally as someone who only encounters one language in childhood. Naturally, there are plenty of other reasons someone might suddenly find themselves needing to learn a different language. People may immigrate for work or fall in love with someone who speaks a different language. The EU's Erasmus Study Abroad programme has reportedly resulted in so many international romantic relationships that more than 1 million "Erasmus Babies" have been born to international couples who met while studying abroad (Brandenburg et al., 2014; European Commission, 2019). It is likely that these children will grow up bilingual.

However, not everyone who can speak a second language does so happily. Our use of language always reflects the socio-political context we are surrounded by. Historically, some languages have been framed as inferior to others, often due to the cultures and nationalities they are associated with. The present work aims to highlight some of the effects of such social, political, and economic power imbalances on research that investigates the effects of speaking more than one language on non-linguistic cognitive processes. More specifically, we will look at research on intelligence conducted in the first half of the twentieth century (Chaps. 2–4), followed by a brief look at meta-cognition and bilingualism (Chap. 5), before turning our attention to more recent research on bilingualism and executive functioning (Chap. 6).

However, before we turn to the research on these topics, I would like to introduce some examples of power imbalances between languages and how they are reflected in education systems. In general, education systems are designed to give young people the best possible start in life and on the job market. As such, they often reflect cultural, socio-economic, and political views and goals. We can find many different examples of power imbalances between languages in multilingual societies, for example Afrikaans vs English in South Africa, Cantonese vs Mandarin in Hong Kong, or Arabic vs French in Algeria. The following examples look at Welsh vs English, Gaelic vs English, and French vs English. They were chosen as we will be revisiting these locations and their contemporary language landscape in Chap. 2, but it should go without saying that these issues can be observed across the globe.

1.1 Suppressed Languages and Education

Language plays a crucial role in our social, political, economic, and cultural lives. For example, it makes intuitive sense that if we do not share a language with another person, our communication will be severely limited. In cases in which people grow up speaking multiple languages, they may associate these languages with different cultures and consider it important to continue using them to carry on traditions. The fact that the language we choose to use in any given moment carries a substantial amount of meaning to our surroundings has its benefits and disadvantages. For example, speaking Irish had been suppressed in Ireland for several centuries, with some degree of violence and aggression. This was primarily due to socio-political reasons, during the British rule of Ireland, and led to Irish officially becoming a minority language by the nineteenth century. When the Irish Free State was founded in 1921, Irish culture, history, and language were made priorities in the educational plan yet very few parents and teachers welcomed the Irish language plan (Buachalla, 1984). The number of pupils studying in Irish steadily declined from the 1940s onwards, as English was seen as the more versatile language, considering it was linked to more social and occupational opportunities. There is also evidence to suggest that children and adolescents struggled to learn Irish, which in turn led to lower educational attainment (MacNamara, 1966). Meanwhile, Anglophone parents in French-speaking areas of Canada, who had become increasingly frustrated with their inability to speak French, developed, together with local researchers, a French immersion education programme that continues to be both effective and popular today, in Francophone and Anglophone areas alike (Lambert & Tucker, 1972; Turcotte, 2019). In both of these cases, the aim was to promote the use of a (regional) minority language, yet the circumstances in which these efforts were made varied widely.

These differences in how the teaching of minority languages is integrated into education continue to be relevant. For example, in early 2020, the Scottish Conservative education spokesperson, Liz Smith, described the wider introduction of Gaelic as primary language of instruction at schools on the Scottish Western Isles, where Gaelic is still widely spoken, as a "deeply troubling" move that "could put children in the Western Isles at a distinct disadvantage to their peers" (Ross, 2020). It is worth noting that the children are still introduced to English, similar to children in, for example, Canadian French-English immersion programmes. These approaches are not generally linked to negative outcomes in terms of educational attainment. However, poorly executed second-language education, as described by MacNamara (1966), can and has, of course, in the past affected educational attainment. The echo of these pedagogical approaches, along with that of research linking bilingualism to "mental retardation" or "mental confusion", can at times lead to overly strong objections towards bilingual education.

However, the form in which Gaelic-medium education is implemented on the Western Isles is linked to good educational outcomes, suggesting that the children will not be at an academic disadvantage (Conger, 2010; Fránquiz, 2012; Močinić, 2011). Indeed, studies that investigated the effect of Gaelic-medium education on children's educational attainment found that they performed at the expected level across a range of different subjects (O'Hanlon, 2010; O'Hanlon et al., 2010, 2013). Students who were predominantly taught in Gaelic at school were also found to be more likely to enter further and/or higher education (Dunmore, 2014). It seems highly unlikely that this particular form of bilingual education would have any detrimental effects on educational attainment. An education spokesperson can reasonably be expected to know at least some of this information and have a basic understanding of bilingual education. Liz Smith was heavily criticised for her statement and later apologised via Twitter, stating that "[her] concerns did not relate to the quality of GME teaching & learning, both of which have such a strong record" (Smith, 2020). This implies that her primary fear in regard to the switch to Gaelic-medium education on the Western Isles was not about the quality of the lessons, but about the choice of the first language of instruction. Gaelic, as a minority language, holds less power than English as the dominant language in Scotland and England. The implicit concern appears to be a scenario in which children have to become bilingual in order to gain access to opportunities that are only available for speakers of the dominant language.

The social aspects of language come through particularly strongly when education is considered. When a Welsh school that previously taught lessons in English informed parents that they would be switching to Welsh following a council vote, there was an outcry in the village where the school is located (Tickle & Morris, 2017). In this example, it does sound like a somewhat rushed decision, as the language of instruction would also be changed for students in higher school years. Immersion education often introduces the second language early on, partly for reasons of language development and partly because the content of lessons becomes more complex with every year, which can make it more difficult for children to actually immerse in a language. There are good reasons for why most of us begin to learn a new language by using phrases like "Please hand me the pink pen" instead of starting out by writing an essay on the Second World War in a language we barely speak. The more complex and abstract a concept, the more difficult it becomes to explain. In contrast, if a child asks for a pink pen, we can provide feedback by handing them the pink one or we can explain that the one they are pointing at is indeed blue and they need to clarify which one they want. This kind of direct feedback is not available if we try to communicate more abstract thoughts, which makes it more difficult to acquire a language in a context in which we are expected to communicate complex and abstract information.

What stands out from the reporting of the case, however, is that people who live in a bilingual country, and who were previously perfectly happy to have a monolingual school, were suddenly angry at the prospect of having a monolingual school that uses the *other* official language of that country, with one Labour MP reportedly going so far as to compare the move to Welsh-medium education to apartheid (Ticke & Morris, 2011). Of course, the change in mood was in large part due to the direction of the change, away from the majority language and towards the minority language. Living in a country with two official languages does not equate to being able to speak these two languages, and the social, political, cultural and economic implications of this have influenced both research and policy making throughout history. However, before we take a closer look at this history, let us briefly discuss what bilingualism actually means.

1.2 Who Is Bilingual and Who Is Monolingual?

When we discuss the effects of bilingualism, it is important to remember that not all bilinguals are alike and that this has implications for researchers and how they design studies. It is, for example, not uncommon for people to pick up a few words in a foreign language. Perhaps they have been watching a TV show that incorporates phrases from other languages or they prepared for a trip abroad, hoping to order their meal in the local language. However, this does not necessarily make them bilingual. There is a spectrum that ranges from being monolingual to being fully fluent in more than one language, and many people fall somewhere in between these two extremes. Additionally, there is the question of how to use the word "bilingual". Much of the research interested in the effects of bilingualism on non-linguistic cognitive skills uses "bilingualism" as an umbrella term that includes all multilinguals (Gathercole et al., 2014; Morales et al., 2013; Salvatierra, & Rosselli, 2011). The "bilinguals" in these studies may speak only two languages or they may speak six or more. Information on any additional languages spoken by bilingual participants is often not provided, and even if it is mentioned that bilinguals spoke more than two languages, they are often referred to as bilinguals regardless of this (DeLuca et al., 2019). Some of the more recent studies on the effect of bilingualism on executive functioning have made a distinction between bilinguals, trilinguals, and/or multilinguals and have reported differences between these groups (Poarch & Bialystok, 2015; Poarch & van Hell, 2012). These will be discussed in greater detail in Chap. 6. For the purposes of the present work, I will adhere to the convention of using "bilingual" to describe anyone who speaks two or more languages, unless the research I refer to distinguishes this group further (e.g. bilinguals and trilinguals). Besides this being the convention for contemporary research, it is worth noting that, historically, "bilingualism" has often been estimated through very unreliable methods. For example, someone with an Italian-sounding last name living in the United States may have been assumed to speak Italian and English, regardless of whether this was actually the case. Thus, for historic research in particular, it is difficult to determine if the "bilinguals" spoke one, two, or more languages retroactively.

Besides number of languages spoken, there are several other factors that can be used to distinguish between different types of bilinguals. "Bilingual usage" is a term that describes how frequently and in what contexts bilinguals use the languages they speak. In general, we can distinguish "balanced bilinguals", who speak both languages roughly equally often, and "non-balanced bilinguals", who use one language more frequently than the other. This is often the case if someone moves abroad on their own, either permanently or temporarily. As an example, consider a French native speaker, who was born and raised in France and learned English as a second language. If that person were to move to New Zealand, it is likely that they would be expected to speak English at work and that most of their local social circle would also communicate in English. Their opportunities to speak or write in French would be very limited compared to the opportunities in France, where English would very likely have been the non-dominant language in their life. However, a French-English bilingual who was born and raised in the French-English bilingual Quebec would frequently have the opportunity to use both languages, depending on circumstances and their own preferences. It is worth noting that these are only examples and that not everyone who moves abroad spends their workdays speaking the local lingua franca, just as not everyone who lives in a bilingual region speaks both languages fluently.

This brings us to "fluency" or "proficiency". Very few of us would consider ourselves fluent in a language after learning a couple of phrases and key words to order a pint in the language of the country we intend to holiday in. During that week or fortnight, we might use these phrases quite frequently but that, of course, does not equal being able to speak the language proficiently. When it comes to proficiency, we can distinguish between high-proficient bilinguals, who are fully fluent in at least two languages, and low-proficient bilinguals, who are fully fluent in one language but whose proficiency in the other language(s) is lower. Someone who grows up in a bilingual society may be exposed to the same language at home and at school but only occasionally encounter a second commonly spoken local language. For example, a Welsh child who speaks English at home and at school may pick up some basic language skills in Welsh but may never become fully proficient in Welsh. This is also the reason why the current recommendation to raise children bilingually is to speak the less well-represented language at home. As an example, if Italian parents living in Brazil intend to raise their children as Italian-Portuguese bilinguals, it would be recommended that they speak Italian with their children. At the same time, if a child learns to speak a language this does not necessarily mean they can use it in other contexts, for example in writing. Similarly, if someone learns to write a language, they may not be exposed to the spoken language and become more proficient in it in writing, particularly if their first and second languages rely on different alphabets (e.g. Latin and Cyrillic). However, it is notable that this distinction was, and continues to be, rarely made in research.

Besides the frequency of use and how proficient bilinguals become, it is also important to consider at what age people acquired a language. Most of us learn to use a language, signed or spoken, by soaking in information about it from the moment we are born. In some cases, this includes more than one language, and if these children continue to be exposed to both languages, they will most likely become bilingual. People who have started to use more than one language in childhood are called early bilinguals. Researchers seem unable to agree on one specific age that could be used as a cut-off point to distinguish between early and late bilinguals; however, the general consensus is that early bilinguals learn to speak two or more languages prior to puberty, with the cut-off age around seven to 10 years (Luk et al., 2011; Moradi, 2014; Pelham & Abrams, 2014). Some early bilinguals will be exposed to two or more languages from birth (i.e. "simultaneous" or "parallel" bilinguals) while others learn to speak one language first and acquire a second one at a later point in childhood (i.e. "sequential" bilinguals). It is worth noting that our ability to master different aspects of language appears to decline at different ages. For example, our ability to acquire syntax and grammatical rules only declines in mid- to late adolescence while the ability to acquire native-like pronunciation in a language declines drastically around five years of age (Birdsong, 2018; Granena & Long, 2013; Hartshorne et al., 2018). Thus, the chosen cut-off age has different implications for different aspects of language.

Naturally, every child develops differently and thus it is likely that the developmental changes that can have a negative effect on language acquisition in late bilinguals do not occur at the same age for every child. Late bilinguals acquire a second language in adolescence or adulthood and, on average, are more likely to have a noticeable accent, are less likely to become fully proficient in their second language, and generally seem to find it more difficult to learn a second language (Flege et al., 1999; Granfeldt, 2018; Piske et al., 2001). Of course, this may not be the case for every person and, occasionally, someone who learned a language at a later stage in life will achieve the same fluency and pronunciation as a native speaker, but this tends to be the exception. At the other end of the spectrum, people who grew up speaking two languages may experience "language attrition" for the language they use less frequently as an adult. Language attrition expresses itself differently in different people, and the "pairing" of languages can have an effect on its severity. People may realise that their sentences have become less complex and they now rely on simpler syntactic structures. Words may be more difficult to remember, as the speaker's active vocabulary "shrinks", or they might struggle with the pronunciation of some words or phrases that they previously used with ease. While it is possible for someone to experience language attrition to such an extent that they become unable to speak a language they were previously fluent in, most people retain at least some basic skills in the unused language (Köpke et al., 2007).

Fluency in a language is often described as "functionally proficient" or "functionally fluent". Both of these are incredible useful and widely applied terms, yet they are very rarely clearly defined (e.g. Bialystok et al., 2004; Colzato et al., 2008; Costa et al., 2008). For the purposes of this book, the phrase "functionally proficient", or "functionally fluent", will refer to someone who can hold a basic conversation in a language and who uses mostly appropriate vocabulary. Their syntax and grammar may not be perfect but a native speaker would be able to discern what the speaker is trying to communicate. Additionally, our imaginary bilingual person would be able to understand a native speaker, if the native speaker spoke clearly and in simple words.

It should be noted that the terminology of "native speaker" vs "non-native speaker" is viewed as problematic by some (Jenkins, 2017). A large part of the concern in regard to this stems from discrimination against non-native speakers in occupational contexts, in particular for language teachers (Mahboob & Golden, 2013). However, another aspect of the problem is that people oppose the "non" in "non-native", as it is perceived to define someone through something that they are missing rather than something positive, that is, the ability to speak the language in question (Jenkins, 2017). Similar concerns have been raised in regard to other terminology, such as "mother tongue", which is seen to place too much emphasis on the languages a person is exposed to in early childhood and thus likely to learn. Naturally, the term may also be problematic if a child is born into a family with fewer or more "mother tongues" than precisely one. This could be, for example, because it is raised by two mothers who speak different languages with the child, the child-bearing person does not identify as female or mother, or the child is born into a gay family. It may be someone's instinct to use "first language", "second language", and so on instead but there we face the issue of identifying in which chronological order someone learned languages. A child born into a bilingual family is likely to acquire both languages simultaneously, which would mean it has two "first languages". This overlap in chronology can also happen at later points in life, for example because people study two languages in parallel. Similarly, someone may start to take language classes, stop, and later continue. If they start to learn another language during that break, which one would be the second vs third language? Another alternative suggestion has been made by Hackert (2012), who proposes that the terms "native/non-native users" should be used instead. However, as much of the concern in regard to terminology stems from the "non" in non-native, this suggestion is unlikely to fully address this issue (Jenkins, 2017). For this reason, the present publication will continue the current convention of using the terminology of native and non-native speaker.

To provide a frame of reference for what language a native speaker with low linguistic ability would be expected to understand, many countries now provide information in "simple" language. For example, the *Life in the UK* textbook that accompanies the immigration test of the same name in the United Kingdom is written in "simple English", while the German job centre ("Bundesagentur für Arbeit") offers a version of their website in "leichter Sprache" (i.e. "easy language"; Arbeitsagentur (n.d.)). This is partly done for the benefit of low-proficient immigrants who may be engaging with these outlets, but it also allows native speakers who may struggle to process more complicated language to access this information. This could, for example, include someone with cognitive deficits or special educational needs, or someone who simply prefers to use an easier and more accessible language. In other cases, someone may come from a less privileged background and/or may not have received adequate education, which can make it difficult to fully comprehend more complex information.

However, this also highlights a double standard. If a bilingual non-native speaker speaks a language and their pronunciation differs from the standard pronunciation, or their grammatical choices are inaccurate, we often relate these factors to how fluent they are in that language (e.g. Andreou & Karapetsas, 2004). If a monolingual native speaker violates grammatical rules or applies a non-standard pronunciation, for example because of a dialect, no one considers them not fully fluent in the only language they speak. Electrophysiological research even suggests that we process semantic and syntactic errors differently if they are made by a native speaker compared to when they are made by a speaker with a non-native accent. If someone who sounds like a native speaker makes such a mistake, we process it as linguistic error, but this does not appear to be the case if the same mistake is made by someone who speaks with a non-native accent (Gibson et al., 2017; Hanulíková et al., 2012). It is likely that the monolingual in this situation would be described as having "low linguistic ability" but we would not describe them as "without language" (e.g. Hunt et al., 1975). Of course, some children remain mostly non-verbal throughout childhood and may never acquire language, which is usually linked to developmental delays and cognitive deficits (Bishop, 1994; Botting, 2005). However, in healthy developed adults no one would conclude that they "did not speak the language" due to transgressions from the standard dialect or grammatical errors.

Even other native speakers may struggle to understand someone whose speech deviates from the expected standard, for example received pronunciation in English, Hochdeutsch in German ("high German"), or the современный русский литературный язык in Russian ("modern Russian literary language"). What all of these standard pronunciations have in common, however, is that they themselves are dialects, but they happen to be the ones that we now teach in schools, report the news in, and expect to see books written in. In most cases, there is more than one reason why a dialect becomes the standard dialect that is representative of a language. However, factors that appear to benefit it are prestige, codification, and being spoken across a wide geographic region (Huesmann, 1998). Indeed, the standard version of a language is often that spoken by the most dominant and influential social classes (Bourdieu, 1991). As a consequence, other social classes may struggle to adapt to the standard version of a language, which comes with serious implications in regard to educational achievements, as well as social and occupational opportunities. Dialects can also form part of our social identities, reflecting what groups we feel we belong to or may feel excluded by (Edwards, 2017). For example, Brady (2015) studied teenagers who found it easier to express themselves in non-standard English when speaking with their peers. However, at the same time many of them reported sanctions for the use of non-standard English in school lessons, which highlights that the standard dialect is often held in higher regard than non-standard dialects. Even if monolingual native speakers are able to apply standard pronunciation to their speech, many will still speak a second dialect at home with their family and friends. Others might grow up without ever fully embracing the standard dialect but move to a different area of the country where they pick up another non-standard dialect, requiring them to switch back and forth, depending on who they speak with and the context of the conversation. Additionally, people use dialects to actively separate themselves from other social groups. In the case of the teenagers studied by Brady, they indicated that part of the appeal of using non-standard language was that it set them apart from the adults in their lives.

At this stage it is perhaps the right time to stress the difference between an accent, a dialect, and a language. If someone speaks with an accent, it affects their pronunciation only. In contrast, a dialect comes with its own set of grammatical rules and dialect-specific vocabulary. A dialect can progress into a language in its own right, although the point at which a dialect becomes a language remains heavily debated. For example, while some researchers consider Cantonese and Mandarin dialects of Chinese, others treat them as two separate languages (Chen et al., 2004; Forsyth et al., 2007). A rough rule of thumb is to look at the "linguistic distance" between two dialects or languages. In other words, we can assess the differences among the pronunciation, grammar, and vocabulary of two linguistic systems. If speakers of two dialects understand one another, we can assume mutual intelligibility (or "mutual comprehensibility") of both linguistic systems and would continue to consider them dialects. However, if the two linguistic systems are not mutually intelligible, they would generally be considered to be two different languages. Naturally, there are limits to the applicability of this rule. For example, speakers of Slovak and Czech are generally able to understand one another well but each language has its own standard form and linguistic system. Additionally, socio-linguistic aspects will always have an influence, on some level, on how readily we are willing to view an established dialect as a language. In the case of Slovak and Czech, both languages are linked to different countries, which may lead us to accept them as two separate languages more readily than if both languages were equally represented in the same country.

The implication of this is that the monolinguals in one study might spend most of their time speaking the standard dialect of their language and have very little knowledge, if any, of a foreign language, while the monolingual participants in another study frequently switch between dialects and remember a surprising amount from those four years they studied French at school. Similarly, the bilingual participants in one research project may be highly proficient early bilinguals and live in a bilingual society, while the bilinguals included in another research project rarely speak their second language and are only functionally proficient. It is not clear what effect smaller differences in the degree to which someone is monolingual or bilingual have on research that studies potential effects of bilingualism on non-linguistic cognitive functions. However, what you will notice as we look at the progression of research in this area, perhaps with some surprise, is that relatively few researchers address the implications of these differences for their own participants or when they draw comparisons across different studies.

1.3 Unimodal and Bimodal Bilingualism

Often, when we describe someone as bilingual, it is automatically assumed that that the person in question speaks two languages. However, this assumption excludes many people who sign and speak a language, or who sign two languages. Sign languages have their own vocabulary, set of grammatical rules, and local dialects. The main difference between signed and spoken languages is the modality used to communicate, which is either through spoken words or visual signs. If someone speaks one language and signs another language, they are considered to be a "bimodal bilingual". In contrast, if a person speaks two languages or signs two languages, they are considered "unimodal bilinguals". It is worth remembering that not every person who uses a sign language is deaf; many are able to hear perfectly well. One readily apparent difference between bimodal and unimodal bilingualism is that bimodal bilinguals can use both languages at the same time, for example a person can sign in British Sign Language (BSL) and speak English while doing so. This is not possible for unimodal bilingual signers or speakers. The implications of this are that the cognitive processes and cortical correlates of bimodal and unimodal bilingualism are different (Mineiro et al., 2014). In other words, bimodal bilinguals rely on different mental processes and brain areas to comprehend and produce spoken languages and sign languages. Admittedly, there is some overlap between how sign languages and spoken language are processed by bimodal bilinguals, but this is limited. Traditionally, there has been little research on bimodal bilingualism. Instead, researchers have focused on testing members of the deaf community on psychometric measures, particularly intelligence, with little interest in bimodal bilingualism or unimodal signed bilingualism (e.g. American Sign Language and British Sign Language). Historically, studies on sign languages do not appear to have considered bilingualism as part of the research design, although this changed towards the end of the twentieth century (Vernon, 2005). Later research frequently referred to the differences between bimodal and unimodal bilingualism to highlight limitations of findings, but there is little information of this kind available for earlier studies on the topic (e.g. Emmorey et al., 2016; Olulade et al., 2015).

The history of research with people who use sign language is inadvertently linked to research with deaf signers and subject to the socio-political situation of the deaf community over time. While this history is incredibly interesting, although at times depressing, covering it is beyond the scope of the present work. The following chapters will discuss research related to unimodal bilinguals who speak two or more languages. It is, however, important to note that it is often not recorded and/or reported whether the people included in the discussed research are also able to sign a language. While more recent research appears to be more likely to report the ability to sign, sign languages are rarely considered in the description of the participants included in a study. In part, this issue arises because the speaking community often intrinsically assumes that languages are spoken and unintentionally phrases questions in a way that excludes sign languages. One recent example, the "Language and Social Background Questionnaire" (LSBQ; Anderson et al., 2018), one of the most extensive questionnaires available to assess how bilingual someone is, asks participants to "List all the language and dialects you can speak and understand including English, in order of fluency". Signed languages are implicitly excluded from the questionnaire by using the word "speak" instead of, for example, "use" or "speak or sign". This reflects an inappropriate bias in research but it also adds another variable that we cannot account for when we compare bilinguals and monolinguals, or try to compare conflicting findings from different studies. For all we know, a group of participants labelled "monolingual" may be a group of fully fluent bimodal bilinguals—it is impossible to determine this if that information is not reported or recorded as part of the investigation in the first place.

What all of this shows is that there are many different background factors that can affect how bilingual or monolingual a person is. Some monolinguals may frequently switch between dialects and may have retained a surprisingly large vocabulary from studying a second language in school. Other monolinguals may stick to the local dialect and may never have studied a second language. Bilinguals, on the other side, tend to range somewhere between being just about able to functionally speak two languages and being fully fluent in several languages, possibly switching between dialects within these languages, and/or being able to sign a language. Whether a person is bilingual or monolingual is not a categorical decision. Instead, people tend to be somewhere between fully monolingual and fully bilingual.

1.4 THE BILINGUAL MIND

"Non-linguistic cognition" is an umbrella term that refers to a range of abilities and characteristics, including intelligence, attention, and memory skills. It can generally be assumed to encompass all mental processes that are not linked to language and are hence "non-linguistic". On the surface level, the inherently linguistic characteristic of being bilingual may seem unlikely to affect non-linguistic cognition. However, even prior to psycholinguistic research on how languages are processed in our minds and brains, people often worried that the strain of speaking more than two languages might lead to some degree of "mental confusion" or affect educational attainment (e.g. Dawes, 1902).

There might be differences in how fluent bilinguals are and how frequently they switch between languages, but all bilinguals can switch between languages, often with very little apparent effort. There is some evidence that suggests a slight delay and an increased rate of inaccurate responses if bilinguals are required to switch languages to perform a task, compared to when they perform the same task without having to switch languages (Green, 1998). Part of the reason why bilinguals are able to switch between languages with relative ease is that they do not selectively "switch on" or "switch off" a language. Both languages are constantly active in the bilingual mind, with access to vocabulary, grammatical rules, and pronunciation information. We know this based on research with interlingual "cognates" and "homonyms", the conclusions of which have also been supported by brain imaging and electrophysiological data.

Interlingual cognates are words that look roughly the same, sound roughly the same, and mean roughly the same in two languages. For example, the English word "create" and the Spanish word "crear" would be considered interlingual cognates. In contrast, interlingual homonyms, or "false friends", appear to be the same but mean something different in each language. They come in two forms, homographs and homophones. Homographs are words that are spelled the same in both languages but mean something different in each language. For example, the German word for mobile phone is "handy", and while mobile phones are handy to have, it is not quite the same as the English word "handy". Homophones, as the name suggests, sound the same but have a different meaning in each language. The Russian word for "(corner) shop", for example, is "магазин" (magasin) and is pronounced similar to the English word "magazine" but of course has a very different meaning.

Interlingual cognates and homonyms can be used in lexical decision tasks, which require participants to judge words as belonging to a specific category. For example, one widely used version of the lexical decision task asks participants to distinguish between real words and made up words (also called "non-words") presented on a computer screen (e.g. Yap et al., 2015). Varying what words are used for the task, for example how frequently the word is generally encountered in daily life, how long it is, or what context it is associated with, allows us to gain some insights into how language is represented in our minds. In the case of bilingualism, we can, for example, ask bilinguals to identify which word belongs to which language, as did Dijkstra et al. (1999). They conducted a series of experiments with Dutch-English bilinguals and found that interlingual cognates were categorised faster than neutral control words or non-words. These findings were later replicated by Lemhöfer and Dijkstra (2004), who also reported evidence that contextual cues may facilitate recognition of words in one language over the other in bilinguals. Homonyms, in contrast, delay categorisation. It is assumed that this is the case because the meaning of the word is "activated" in both languages. As the semantic information matches in both languages for cognates, participants respond faster to them than to other words. The opposite is the case for homonyms; bilinguals need to resolve the conflict between competing semantic information from both languages in order to categorise the word appropriately, which delays response times (see Dijkstra, 2007 for a review). This conclusion is supported by findings from electrophysiological studies, which linked activation measured by an electroencephalogram (EEG) to behavioural measures. For example, Thierry and Wu (2007) reported findings that suggest that bilinguals may subconsciously translate words, even if the translation is not relevant to the task they are performing.

There is some evidence that implies that the frequency with which bilinguals switch between languages may affect the strength of the facilitation effect of cognates and the conflict effect of homonyms (Jared & Szucs, 2002). Both effects appear to be stronger if bilinguals recently switched between languages, which suggests that the frequency with which languages are switched, and whether bilinguals recently switched between languages, may affect how "active" both languages are. While both languages can be assumed to be active in bilinguals, it is not clear how the representation of different languages is balanced in the bilingual mind, especially as this is likely to be highly dependent on the individual and their circumstances. Bilinguals who speak two languages with a small linguistic distance, for example Dutch and German, will also encounter interlingual cognates more frequently than bilinguals who speak two languages that are more distant, for example Dutch and Vietnamese (Schepens et al., 2013). This further emphasises that different bilinguals may experience being bilingual differently in terms of cognitive processes.

Eye-tracking studies provide further support for the parallel activation of several languages in bilinguals. These studies (e.g. Marian & Spivey, 2003a, 2003b) allow us to assess between-language competition and within-language competition. Their general set-up presents participants with several different objects. One of these objects will then be named and participants are expected to direct their gaze to it. The target object is always phonologically similar (e.g. pan vs pen) to one of the other presented objects but the language in which the two objects are phonologically similar varies from testing condition to testing condition. For example, if the task is performed in English with Dutch-English bilinguals, task instructions are given in English. In that case, if participants have to look at a "bin" the phonetically similar word might be "tin" in English. Performance on this task version is indicative of within-language competition between the words, and participants are expected to look at the picture of a bin. It is furthermore expected that they are more likely to look at the picture of a "tin" than pictures of other distractor items, that is, those that are not phonetically similar to "bin".

In another variation on this task, the distractor item is phonetically similar in the other language spoken by the bilingual participant, and a Dutch-English bilingual instructed to look at a "broom" might find themselves drawn to look at the picture of a tree instead, as the Dutch word for tree is "boom". Task performance in this condition is viewed as indicative of between-language competition, as two words from two different languages compete for representation. Finally, the phonetically similar distractor item can be phonetically similar in both languages. For example, if asked to look at a "hook" the picture of a book could represent the Dutch and English word for it, "boek" and "book" respectively, both of which are phonetically similar to "hook". Trials of this type would be considered to provide an indication of simultaneous competition of words from both languages.

Bilinguals have repeatedly been found to experience both within-language and between-language competition, including Spanish-English bilinguals (Blumenfeld & Marian, 2011), Dutch-English bilinguals (Schulpen et al., 2003; Weber & Cutler, 2004), and Russian-English bilinguals (Kaushanskaya & Marian, 2007). However, whether between-language competition is observed depends, to some degree, on the environment in which bilingual participants live. If someone fluent in two languages is primarily surrounded by their first language and primarily speaks that first language, they are less likely to experience strong between-language competition (Marian & Spivey, 2003a, 2003b; Spivey & Marian, 1999). This highlights something very important: while we can assume with relative certainty that all languages spoken by someone are "active" and can be accessed, this does not mean that all languages are equally accessible at all times. It is likely that languages that are used more frequently are "more active", that is, that linguistic knowledge can be accessed with more ease than the linguistic knowledge linked to a language that is less frequently used. It is not entirely clear how this activation of different languages is balanced, however, and even if we could put a number on this, it is likely that it would be a highly individual number. What we do know is that bilinguals who are immersed in an environment in which their second language is the more dominant one appear to have reduced access to the vocabulary and grammatical rules of their first language (Linck et al., 2009).

While this kind of research provides insights into how bilinguals comprehend languages, other studies have focused on how bilingualism may benefit language production, that is, producing speech. Interlingual cognates do not only facilitate language comprehension, they also benefit language production, for example in resolving tip-of-the-tongue states (Costa et al., 2005; Gollan & Acenas, 2004; Kohnert, 2004). It is assumed that the reason for this is that while someone may struggle to remember the word for, for example, cat, in English, someone bilingual can still access this word in their other language, which subsequently benefits the retrieval of the word they are looking for in the language relevant to the context they find themselves in. Additionally, EEG studies suggest that the orthographic neighbourhood of words does not distinguish between languages. Orthographic neighbours are words that are similar to one another; for example, "rat", "fat", "cat", and "chat" would be considered orthographic neighbours. Using EEG, we can record an event-related potential (ERP) called N400. ERPs are small voltages elicited by brain structures; the N in N400 stands for negative and the 400 for 400 ms. It is thus perhaps not surprising that the N400 shows as a negative trough on an EEG recording and occurs roughly 400 ms (approx. 300–600 ms; Kutas & Hillyard, 1980) after presentation of the stimulus we are interested in.

Importantly, the N400 ERP is sensitive to the size of words' orthographic neighbourhood, which means it is bigger if a word is linked to more similar words (Holcomb et al., 2002). This means that if both languages are active in parallel in bilinguals, words that have a larger orthographic neighbourhood across languages should result in a larger negative amplitude. When Midgley, Holcomb, Walter, and Grainger (2008) tested this hypothesis, this is precisely what they found: words with larger cross-language orthographic neighbourhoods elicited larger N400 amplitudes, adding further support to the notion that both languages are active in bilinguals at all times. The same pattern of an increased negative N400 amplitude is observed when bilinguals produce interlingual homophones, which again shows that both languages are active in parallel (Christoffels et al., 2016).

Helpfully, the N400 is also sensitive to unexpected sentence endings (Kutas & Hillyard, 1980). For example, "She ate a sandwich" would be an expected sentence ending, as it is not unusual to eat a sandwich. In contrast, "She ate a bicycle" would be unexpected, as we generally do not eat bicycles. Fitzpatrick and Indefrey (2010) used the N400s sensitivity to unexpected sentence endings to test cross-language activation in Dutch-English bilinguals. Their participants listened to sentences in English and these could either end with a word that made sense in English or with a word that did not make sense in English and Dutch. As a third option, the last word in the sentence could start with a sound (i.e. phoneme) similar to a word that would be fitting for the end of the sentence, but the actual word would not make sense within the provided context. For example, "He drank his co-", where "co-" could lead to several different words, including cola, Coke, coal, or coat. For the third condition, participants might have heard the sentence "He drank his coat", in which case the unspoken congruent word, that is, the one that would make sense within the context of this sentence, would be cola or Coke. If both languages are active at the same time, incongruent words, that is, words that do not make sense within the context of the sentence, would be expected to affect the N400 regardless of whether they are Dutch or English, which is precisely what Fitzpatrick and Indefrey found. Their data also showed that the N400 occurred later following English words than following Dutch words. Dutch was the native language of participants in this study. This difference in time thus implies that integrating first-language words into meaningful second-language sentences may be more effortful. In summary, we can conclude with some degree of certainty that both languages spoken by a bilingual are active at all times. However, how balanced that "activation" is appears to depend on both the individual and their circumstances.

1.5 Summary of Bilingualism

To summarise, we briefly looked at some examples of how societal and political factors can affect the status of a language and of bilingualism, particularly in the context of education. Languages that offer economic or social advantages are likely to appeal to people, which can increase the chances of someone becoming bilingual and lead to the spread of a language. Similarly, bilingualism may be discouraged or encouraged for political reasons. We also considered different language profiles for bilinguals and monolinguals. For bilinguals, some of the key factors in characterising them are the age at which they became bilingual, how proficient they are in the languages they speak, how balanced they are in the use of their languages, and the similarities between the languages they speak. We may also want to consider the points we can use to characterise language as it is used by monolinguals, such as whether a person switches between dialects or uses grammar correctly. While it is rarely reported, ideally, we would also want to know if a person can sign a language in addition to speaking a language. Another important point to remember as we go forward is that bilinguals cannot "switch off" any of the languages they speak; they always have access to all languages, even though their language skills may deteriorate if they do not use a language for a long period of time.

References

Anderson, J. A., Mak, L., Chahi, A. K., & Bialystok, E. (2018). The language and social background questionnaire: Assessing degree of bilingualism in a diverse population. *Behavior Research Methods, 50*(1), 250–263.

Andreou, G., & Karapetsas, A. (2004). Verbal abilities in low and highly proficient bilinguals. *Journal of Psycholinguistic Research, 33*(5), 357–364.

Arbeitsagentur. (n.d.). *Bundes-Agentur für Arbeit*. Retrieved from https://www.arbeitsagentur.de/leichte-sprache/startseite

Bialystok, E., Craik, F. I., Klein, R., & Viswanathan, M. (2004). Bilingualism, aging, and cognitive control: Evidence from the Simon task. *Psychology and Aging, 19*(2), 290.

Birdsong, D. (2018). Plasticity, variability and age in second language acquisition and bilingualism. *Frontiers in Psychology, 9*, 81.

Bishop, D. V. (1994). Is specific language impairment a valid diagnostic category? Genetic and psycholinguistic evidence. *Philosophical Transactions of the Royal Society of London. Series B: Biological Sciences, 346*(1315), 105–111.

Blumenfeld, H. K., & Marian, V. (2011). Bilingualism influences inhibitory control in auditory comprehension. *Cognition, 118*(2), 245–257. https://doi.org/10.1016/j.cognition.2010.10.012

Botting, N. (2005). Non-verbal cognitive development and language impairment. *Journal of Child Psychology and Psychiatry, 46*(3), 317–326.

Bourdieu, P. (1991). *Language and symbolic power*. Harvard University Press.

Brady, J. (2015). Dialect, power and politics: Standard English and adolescent identities. *Literacy, 49*(3), 149–157.

Brandenburg, U., et al. (2014). *The Erasmus Impact Study. Effects of mobility on the skills and employability of students and the internationalisation of higher education institutions.* Retrieved from https://www.researchgate.net/publication/320267535_The_Erasmus_Impact_Study_Effects_of_mobility_on_the_skills_and_employability_of_students_and_the_internationalisation_of_higher_education_institutions

Buachalla, S. O. (1984). Educational Policy and the Role of the Irish Language from 1831 to 1981. *European Journal of Education, 19*(1), 75.

Chen, X., Anderson, R. C., Li, W., Hao, M., Wu, X., & Shu, H. (2004). Phonological awareness of bilingual and monolingual Chinese children. *Journal of Educational Psychology, 96*(1), 142.

Christoffels, I., Timmer, K., Ganushchak, L., & La Heij, W. (2016). On the production of interlingual homophones: Delayed naming and increased N400. *Language, Cognition and Neuroscience, 31*(5), 628–638.

Colzato, L. S., Bajo, M. T., Van den Wildenberg, W., Paolieri, D., Nieuwenhuis, S., La Heij, W., & Hommel, B. (2008). How does bilingualism improve executive control? A comparison of active and reactive inhibition mechanisms. *Journal of Experimental Psychology: Learning, Memory, and Cognition, 34*(2), 302.

Conger, D. (2010). Does bilingual education interfere with English-language acquisition? *Social Science Quarterly, 91*(4), 1103–1122.

Costa, A., Hernández, M., & Sebastián-Gallés, N. (2008). Bilingualism aids conflict resolution: Evidence from the ANT task. *Cognition, 106*(1), 59–86.

Costa, A., Santesteban, M., & Caño, A. (2005). On the facilitatory effects of cognate words in bilingual speech production. *Brain and Language, 94*(1), 94–103.

Dawes, T. R. (1902). *Bilingual teaching in Belgian schools, being the report on a visit to Belgian schools as Gilchrist travelling student.* Presented to the Court of the University of Wales.

DeLuca, V., Rothman, J., Bialystok, E., & Pliatsikas, C. (2019). Redefining bilingualism as a spectrum of experiences that differentially affects brain structure and function. *Proceedings of the National Academy of Sciences of the United States of America, 116*(15), 7565–7574.

Dijkstra, T. (2007). Task and context effects in bilingual lexical processing. In I. Kecskés & L. Albertazzi (Eds.), *Cognitive aspects of bilingualism*. Springer Science & Business Media.

Dijkstra, T., Grainger, J., & Van Heuven, W. J. (1999). Recognition of cognates and interlingual homographs: The neglected role of phonology. *Journal of Memory and Language, 41*(4), 496–518.

Dunmore, S. (2014). *Bilingual life after school? Language use, ideologies and attitudes among Gaelic-medium educated adults*. (Doctoral dissertation). Retrieved from https://era.ed.ac.uk/bitstream/handle/1842/10636/Dunmore2015.pdf?sequence=2

Edwards, J. (2017). Nonstandard dialect and identity. In R. Bassiouney (Ed.), *Identity and dialect performance: A study of communities and dialects*. Routledge.

Emmorey, K., Giezen, M. R., & Gollan, T. H. (2016). Psycholinguistic, cognitive, and neural implications of bimodal bilingualism. *Bilingualism, 19*(2), 223–242. https://doi.org/10.1017/S1366728915000085

European Commission, Directorate-General for Education, Youth, Sport and Culture. (2019). *Erasmus+ higher education impact study: Final report*. Publications Office. Retrieved from https://data.europa.eu/doi/10.2766/162060

Fitzpatrick, I., & Indefrey, P. (2010). Lexical competition in nonnative speech comprehension. *Journal of Cognitive Neuroscience, 22*(6), 1165–1178.

Flege, J. E., Yeni-Komshian, G. H., & Liu, S. (1999). Age constraints on second-language acquisition. *Journal of Memory and Language, 41*(1), 78–104.

Forsyth, B. H., Kudela, M. S., Levin, K., Lawrence, D., & Willis, G. B. (2007). Methods for translating an English-language survey questionnaire on tobacco use into Mandarin, Cantonese, Korean, and Vietnamese. *Field Methods, 19*(3), 264–283.

Fránquiz, M. E. (2012). Key concepts in bilingual education: Identity texts, cultural citizenship, and humanizing pedagogy. *New England Reading Association Journal, 48*(1), 32–42.

Gathercole, V. C. M., Thomas, E. M., Kennedy, I., Prys, C., Young, N., Viñas-Guasch, N., et al. (2014). Does language dominance affect cognitive performance in bilinguals? Lifespan evidence from preschoolers through older adults on card sorting, Simon, and metalinguistic tasks. *Frontiers in Psychology, 5*, 11.

Gibson, E., Tan, C., Futrell, R., Mahowald, K., Konieczny, L., Hemforth, B., & Fedorenko, E. (2017). Don't underestimate the benefits of being misunderstood. *Psychological Science, 28*(6), 703–712.

Gollan, T. H., & Acenas, L. A. R. (2004). What is a TOT? Cognate and translation effects on tip-of-the-tongue states in Spanish-English and tagalog-English bilinguals. *Journal of Experimental Psychology: Learning, Memory, and Cognition, 30*(1), 246.

Granena, G., & Long, M. H. (2013). Age of onset, length of residence, language aptitude, and ultimate L2 attainment in three linguistic domains. *Second Language Research, 29*(3), 311–343.

Granfeldt, J. (2018). The development of gender in simultaneous and successive bilingual acquisition of French–Evidence for AOA and input effects. *Bilingualism: Language and Cognition, 21*(4), 674–693.

Green, D. W. (1998). Mental control of the bilingual lexico-semantic system. *Bilingualism: Language and Cognition, 1*(2), 67–81.

Hackert, S. (2012). *The emergence of the English native speaker.* Walter de Gruyter.

Hanulíková, A., Van Alphen, P. M., Van Goch, M. M., & Weber, A. (2012). When one person's mistake is another's standard usage: The effect of foreign accent on syntactic processing. *Journal of Cognitive Neuroscience, 24*(4), 878–887.

Hartshorne, J. K., Tenenbaum, J. B., & Pinker, S. (2018). A critical period for second language acquisition: Evidence from 2/3 million English speakers. *Cognition, 177,* 263–277.

Holcomb, P. J., Grainger, J., & O'Rourke, T. (2002). An electrophysiological study of the effects of orthographic neighborhood size on printed word perception. *Journal of Cognitive Neuroscience, 14*(6), 938–950.

Huesmann, A. (1998). Die Standardvarietät. In A. Huesmann (Ed.), *Zwischen Dialekt und Standard: Empirische Untersuchung zur Soziolinguistik des Varietätenspektrums im Deutschen.* Max Niemeyer GmbH & Co KG.

Hunt, E., Lunneborg, C., & Lewis, J. (1975). What does it mean to be high verbal? *Cognitive Psychology, 7*(2), 194–227.

Jared, D., & Szucs, C. (2002). Phonological activation in bilinguals: Evidence from interlingual homograph naming. *Bilingualism: Language and Cognition, 5*(3), 225–239.

Jenkins, S. (2017). The elephant in the room: Discriminatory hiring practices in ELT. *Elt Journal, 71*(3), 373–376.

Kaushanskaya, M., & Marian, V. (2007). Bilingual language processing and interference in bilinguals: Evidence from eye tracking and picture naming. *Language Learning, 57*(1), 119–163.

Kohnert, K. (2004). Cognitive and cognate-based treatments for bilingual aphasia: A case study. *Brain and Language, 91*(3), 294–302.

Köpke, B., Schmid, M. S., Keijzer, M., & Dostert, S. (Eds.). (2007). *Language attrition: Theoretical perspectives.* John Benjamins Publishing.

Kutas, M., & Hillyard, S. A. (1980). Reading senseless sentences: Brain potentials reflect semantic incongruity. *Science, 207*(4427), 203–205.

Lambert, W. E., & Tucker, G. R. (1972). *Bilingual education of children: The St. Lambert experiment.* Newbury House.

Lee Salvatierra, J., & Rosselli, M. (2011). The effect of bilingualism and age on inhibitory control. *International Journal of Bilingualism, 15*(1), 26–37.

Lemhöfer, K., & Dijkstra, T. (2004). Recognizing cognates and interlingual homographs: Effects of code similarity in language-specific and generalized lexical decision. *Memory & Cognition, 32*(4), 533–550.

Linck, J. A., Kroll, J. F., & Sunderman, G. (2009). Losing access to the native language while immersed in a second language evidence for the role of inhibition in second-language learning. *Psychological Science, 20*(12), 1507–1515.

Luk, G., De Sa, E. R. I. C., & Bialystok, E. (2011). Is there a relation between onset age of bilingualism and enhancement of cognitive control? *Bilingualism: Language and Cognition, 14*(4), 588–595.

MacNamara, J. T. (1966). *Bilingualism and Primary Education.* The University Press.

Mahboob, A., & Golden, R. (2013). Looking for native speakers of English: Discrimination in English language teaching job advertisements. *Age, 3*(18), 21.

Marian, V., & Spivey, M. (2003a). Competing activation in bilingual language processing: Within-and between-language competition. *Bilingualism: Language and Cognition, 6*(2), 97–115.

Marian, V., & Spivey, M. (2003b). Bilingual and monolingual processing of competing lexical items. *Applied Psycholinguistics, 24*(2), 173–193.

Midgley, K. J., Holcomb, P. J., Walter, J. B., & Grainger, J. (2008). An electrophysiological investigation of cross-language effects of orthographic neighborhood. *Brain Research, 1246,* 123–135.

Mineiro, A., Nunes, M. V. S., Moita, M., Silva, S., & Castro-Caldas, A. (2014). Bilingualism and bimodal bilingualism in deaf people: A neurolinguistic approach. In M. Marschark, G. Tang, & H. Knoors (Eds.), *Bilingualism and bilingual deaf education* (pp. 187–209). OUP Us.

Močinić, A. (2011). Bilingual education. *Metodički obzori, 13,* 175–182.

Moradi, H. (2014). An investigation through different types of bilingualism. *International Journal of Humanities & Social Sciences, 1,* 108.

Morales, J., Calvo, A., & Bialystok, E. (2013). Working memory development in monolingual and bilingual children. *Journal of Experimental Child Psychology, 114*(2), 187–202.

O'Hanlon, F. (2010). Gaelic-medium primary education in Scotland: Towards a new taxonomy? *Coimhearsnachd na Gàidhlig an-Diugh/Gaelic Communities Today,* 99–116.

O'Hanlon, F., McLeod, W., & Paterson, L. (2010). *Gaelic-medium Education in Scotland: Choice and attainment at the primary and early secondary school stages.* Retrieved from https://www.researchgate.net/profile/Wilson-Mcleod/publication/266883020_Gaelic-medium_Education_in_Scotland_choice_and_attainment_at_the_primary_and_early_secondary_school_stages/

links/568bd1dd08ae8f6ec75233c6/Gaelic-medium-Education-in-Scotland-choice-and-attainment-at-the-primary-and-early-secondary-school-stages.pdf

O'Hanlon, F., Paterson, L., & McLeod, W. (2013). The attainment of pupils in Gaelic-medium primary education in Scotland. *International journal of bilingual education and bilingualism, 16*(6), 707–729.

Olulade, O. A., Jamal, N. I., Koo, D. S., Perfetti, C. A., LaSasso, C., & Eden, G. F. (2015). Neuroanatomical evidence in support of the bilingual advantage theory. *Cerebral Cortex, 26*(7), 3196–3204.

Pelham, S. D., & Abrams, L. (2014). Cognitive advantages and disadvantages in early and late bilinguals. *Journal of Experimental Psychology: Learning, Memory, and Cognition, 40*(2), 313.

Piske, T., MacKay, I. R., & Flege, J. E. (2001). Factors affecting degree of foreign accent in an L2: A review. *Journal of Phonetics, 29*(2), 191–215.

Poarch, G. J., & Bialystok, E. (2015). Bilingualism as a model for multitasking. *Developmental Review, 35*, 113–124.

Poarch, G. J., & van Hell, J. G. (2012). Executive functions and inhibitory control in multilingual children: Evidence from second-language learners, bilinguals, and trilinguals. *Journal of Experimental Child Psychology, 113*(4), 535–551.

Ross, S. (2020). MSPs demand apology for 'highly offensive' Tory comments on Gaelic education. *The Scotsman*. Retrieved from: https://www.scotsman.com/news/politics/msps-demandapology-highly-offensive-tory-comments-gaelic-education-1396281

Schepens, J., Dijkstra, T., Grootjen, F., & Van Heuven, W. J. (2013). Cross-language distributions of high frequency and phonetically similar cognates. *PLoS One, 8*(5), e63006.

Schulpen, B., Dijkstra, T., Schriefers, H. J., & Hasper, M. (2003). Recognition of interlingual homophones in bilingual auditory word recognition. *Journal of Experimental Psychology: Human Perception and Performance, 29*(6), 1155.

Smith, L. [@liz4PSK]. (2020, January 25). *May I apologise to those members of the Gaelic community who have been offended. My concerns did not relate to the quality of GME teaching & learning, both of which have such a strong record.* [Tweet] Twitter. Retrieved from https://twitter.com/liz4PSK/status/1220956484055584768

Spivey, M. J., & Marian, V. (1999). Cross talk between native and second languages: Partial activation of an irrelevant lexicon. *Psychological Science, 10*(3), 281–284.

Thierry, G., & Wu, Y. J. (2007). Brain potentials reveal unconscious translation during foreign-language comprehension. *Proceedings of the National Academy of Sciences of the United States of America, 104*(30), 12530–12535.

Tickle, L., & Morris, S. (2017). We're told we're anti-Welsh bigots and fascists'–the storm over Welsh-first schooling'. *The Guardian*.

Turcotte, M. (2019). *Results from the 2016 Census: English-French bilingualism among Canadian children and youth*. Statistics Canada: Statistique Canada.

Vernon, M. (2005). Fifty years of research on the intelligence of deaf and hard-of-hearing children: A review of literature and discussion of implications. *Journal of Deaf Studies and Deaf Education, 10*(3), 225–231.

Weber, A., & Cutler, A. (2004). Lexical competition in non-native spoken-word recognition. *Journal of Memory and Language, 50*(1), 1–25.

Yap, M. J., Sibley, D. E., Balota, D. A., Ratcliff, R., & Rueckl, J. (2015). Responding to nonwords in the lexical decision task: Insights from the English Lexicon Project. *Journal of experimental psychology. Learning, Memory, and Cognition, 41*(3), 597–613. https://doi.org/10.1037/xlm0000064

CHAPTER 2

Bilingual Education in the Early Twentieth Century

Abstract This chapter discusses early reports of bilingual education in Belgium, which were compiled with the aim to make an informed decision about bilingual education in Wales. It also provides a brief review of two early case studies of bilingual children, which did not find evidence of negative effects of bilingualism.

Keywords Bilingualism • Education • One parent, one language • Language development

2.1 Early Bilingual Education Reports

When it comes to bilingualism, we could go back in time by a considerable number of years. Before we were able to develop the ability to speak a language, our bodies first had to go through changes that would allow us to use spoken language the way we do today. Our ancestors' skulls started to change in such a way approximately 4.5 millennia ago—and then it still took us another four millennia or so to start producing some form of speech, but it is likely to have had little in common with languages spoken today (Clark & Henneberg, 2017). Considering the time that has passed, it is not surprising that it is difficult to pinpoint individual events, but we can make an informed guess. Estimates of when humans developed language range widely, from approximately 350,000 to 50,000 years ago, with most estimates falling somewhere in between these two extremes (Gell-Mann & Ruhlen, 2011; Perreault & Mathew, 2012). What we can say with relative certainty is that by the Upper Paleolithic (50,000–12,000 BCE), humans had developed the ability to use language (Schreyer, 2002).

One reason why there is such a wide range in estimates is that not everybody agrees on when an "oral communication system" becomes "a language". Similar to the question of when a dialect becomes a language in its own right, we can ask when a system of sounds used to communicate becomes a language, and different people will apply different criteria to answer this question. The other aspect, of course, is that researchers have to rely on fossils and archaeological findings to date when language emerged. By default, this means that, depending on the method used to estimate the emergence of the first human language, different researchers will arrive at different conclusions and can only provide a very rough estimate.

In order for humans to be bilingual, they need to use at least two languages. It is not clear how, specifically, humanity arrived at a point at which people started to speak two distinct languages, but what we do know is that by approximately 3000–2500 BCE, Sumerian-Akkadian bilingualism was quite common in Mesopotamia (Deutscher, 2000; Thureau-Dangin, 1911). One issue with dating the emergence of archaic languages and human bilingualism is that in order for us to find evidence of them, humans needed to develop written communication. However, there is some evidence to suggest that most humans living in the Neolithic (10,000–4500 BCE) were monolingual, although some travelers and traders may have been bilingual (Naccache, 2016). Nevertheless, without physical evidence of bilingualism, its emergence is difficult to date. It is simply not possible to find traces of spoken conversations, and it is likely that as soon as more than one language existed, some individuals became bilingual. For the purposes of this book, we are not going to go quite as far back in time as the Neolithic but will instead focus on the more recent history of bilingualism and the research that investigated whether speaking more than one language may affect non-linguistic cognition.

It is also worth noting that most of the discussed research was produced in Europe and Northern America. This focus on the Western world was not intended when I originally set out to look into the history of research with bilinguals, but it reflects both the dominance of Western (and in particular Anglo-American) research in publications on the topic and the way in which research often depends on networks. If a paper is published in a British journal and written in English, it is more likely to be read by someone familiar with the publication and, of course, requires this someone to have a good understanding of English. While it was not unusual to provide abstracts in more than one language up until the early to mid-twentieth century, this generally only made the research accessible to other speakers of European languages. For example, Barke and Williams (1938) provided an English abstract, matching the English language of the article, as well as a French and German abstract. As mentioned earlier, the choice of a language can invite or exclude someone from a conversation, and the scientific discourse is no exception to this rule. Someone unable to speak any of these languages may be able to ask a colleague who does for a competent translation, but if that option is not available either, they simply will not be able to access this information. Additionally, there were practical considerations limiting access to publications. Prior to the days of the internet, access to research often depended on access to either the researchers or a physical copy of their findings. Both of these would be difficult to obtain if the source of information was a considerable distance away.

As a consequence of these factors, much of the early European work I researched refers to other European work, with some influences from Canadian and American research. It was rare to find reference to a study conducted outside of this realm, although it did occasionally happen (e.g. Yoshioka, 1929). It is, however, important to acknowledge that both Asia and Africa have an incredibly rich history of multilingualism and historic research on multilingualism (e.g. Gxilishe, 1900), but even African and Asian records are often linked to European cultures (e.g. Tarn, 1902). The socio-political status of individual languages and multilingualism was also generally very different in these geographic regions. This was partially due to inherent cultural differences, but it is also important to consider colonialism in this context. As an example, the decolonisation of the British Empire did not begin until the late 1940s, when India was given independence, and the formal decolonisation of the French Empire did not begin until the 1960s (Clayton, 2014; Singh, 1984). The influences of French and English in occupied territories cannot be ignored, and it would be impossible to do them justice within the constraints of the present work.

This chapter will discuss two educational reports that focused on bilingualism from the early 1900s, and two European case studies with bilingual children that were conducted around the turn of the twentieth century. The rationale behind choosing these specific reports is that they formed the basis of crucial work we will discuss in the next chapter (Saer, Smith, & Hughes, 1924). They are also well-documented education reports that illustrate the landscape of European bilingual education at a time when researchers increasingly became interested in how bilingualism affected childhood development. The two educational reports we will look at were commissioned to inform the approach to Welsh-English bilingual education in Wales, and both focused on the Belgian education system to learn more about bilingual education.

The first of these reports was delivered in 1902 by Thomas Dawes, who visited a number of Belgian bilingual schools. The Belgian school system had been subject to rapid changes in the years leading up to Dawes' visit, and French and Flemish[1] had only received equal status four years before the publication of his report, when the Coremans-De Vriendt law was passed (Caron, 2014). While the new law was supposed to establish equal rights for French- and Flemish-speakers, it fell somewhat short of that goal. It considered legal texts composed in Flemish as valid, whereas previously all legal documents had to be composed in French. Parliamentary debates could now take place in French or Flemish, and election papers were now also available in both languages (Hensley, 2008). However, in reality, Flemish-speakers were required to learn French while the French-speakers experienced little need to learn Flemish. Public life had previously been built around the needs of French-speakers and in particular French monolinguals. As such, the societal and political structures still disadvantaged the Flemish-speaking population (see 1897 meeting of the Belgian Parliament [Sénat. Annales Parlementaires, 1897]). This was also reflected in Belgian education at the time. Until 1883, Belgian schools taught exclusively in French but a series of new "laws on the use of language" had gradually given more rights to speakers of Flemish and introduced the option of Flemish-language and bilingual education (Mettewie & Mensel, 2020; van Gerwen et al., 2017). This change came at the end of a heated

---

[1] There is an ongoing debate on whether Flemish is a language in its own right or a dialect of Dutch, which should be acknowledged. There are arguments in favour of both sides of the debate; however, whether people viewed Flemish as a language or as a dialect of Dutch has little impact on how the language spoken by the Flemish population in Belgium was treated in a political, legal, educational, and social context and, as such, the topic will not be discussed in greater detail here. For further reading on the distinction between Dutch and Flemish, see, for example, Deprez (1999) and Kroon et al. (2018).

and several-decade-long debate between Catholic and secular Belgians, titled the 'First School War' (the Second Belgian School War took place in the 1950s; Deschouwer, 2006; Devuyst, 1983). At the end of it, Belgians had a choice between sending their children to secular or Catholic schools, although religious education later became mandatory at all schools in the country.

For his report, Dawes (1902) travelled across Belgium to speak to different Directors of Schools, whose role was similar to that of a headmaster. He was primarily interested in the practical and organisational requirements of establishing a bilingual education system and hoped that the expertise held in Belgium would be able to inform the Welsh-English approach to education in Wales. On the national level, he found that the French-speaking Walloon schools generally outperformed the Flemish-French bilingual schools, and one Director of Schools was quoted as stating that "the pupils are somewhat confused with the two languages, there is a great mental effort in changing from one language to another" (pp. 49–50). Interestingly, Dawes did not investigate this statement further; it is merely mentioned in passing, suggesting that he did not share these concerns. The verdict of his report strongly emphasised the benefits of bilingualism, highlighting the increased social, educational, and professional opportunities.

The second Welsh report on Belgian bilingual education was provided by James Williams (1915), only slightly more than a decade later. In Williams' account, he acknowledged that the Belgian situation was different to the Welsh one, as he perceived French and Flemish as having approximately the same socio-economic and political power, while English dominated life in Wales. This might come as a surprise, considering that in the 1890s, French-speaking Belgians had little reason to learn Flemish, as most official business and day-to-day life could easily be completed as a French monolingual. By the 1910s, Walloons were still not overly keen to learn Flemish, but with increasing pressures for state services to be bilingual, French monolinguals who worked for the government were in a vulnerable position. Moreover, voices emerged that suggested that the Belgian educated elite should indeed be able to speak both Flemish and French. Hensley (2008) cited a German historian from the University of Liège in Belgium, who in 1911 stated that "in Flanders the peasant knows only Flemish, and in Wallonia only French, fine, but those who pretend to be the elite of the nation cannot content themselves with this provincial ideal". Germans had lived in Belgian ever since the Kingdom of Belgium was established in 1830, but they constituted, and continue to constitute, only a small minority of the population of Belgium. Godefroid Kurth, the historian quoted by Hensley, acknowledged that this also put him in the privileged position of being neither part of the Flemish nor Walloon community, which may perhaps have allowed him to view the issue with more emotional distance. As Hensley emphasised, the Walloon elite was not opposed to the study of second languages; they simply preferred to spend the time and effort required to do so on Ancient Greek and Latin. Regardless of these discrepancies in opinion, however, Flemish gained in sociolinguistic status.

In Williams' (1915) opinion, Welsh was generally spoken to communicate private information, for example when talking to friends and family. English, on the other hand, was used to conduct business and was thus often linked to employment and business profits, which disadvantaged Welsh monolinguals. At the end of his report, Williams nevertheless concludes that "it is the plain duty of statesmen and teachers to dispel the permanent cause of disunion among citizens of the country" (p. 104) and recommends strengthening bilingual education. This agrees with Dawes' (1902) final conclusion, that the benefits of bilingual education outweigh any potential negative effects. Notably, Belgium remains a multilingual country today, with English commonly spoken by most Belgians. However, Flemish has now become the more socially dominant language, although official services and education remain available in both Flemish and French (Blommaert, 2011).

Not long after William's report was published, the 1918 Education Act of England and Wales (c. 39) stipulated that local authorities in Wales had to provide information on the local education schemes, including provisions to teach Welsh. Prior to 1918, the provision of Welsh education had been mandatory in a small number of local authorities, and optional in three larger Welsh towns, but there was no centralised system for Welsh-medium education that covered all of Wales (Redknap et al., 2006). The Education Act aimed to establish what the current Welsh-medium teaching provisions were in each local authority and, to some degree, how to extend them to local authorities that did not offer teaching in Welsh to predominantly, or exclusively, Welsh-speaking students. This was an important step towards providing Welsh-medium education in Wales, and was marked by an increased interest in the potential disadvantages and benefits of bilingual education.

2.2 Early Case Studies of Bilingual Children

There are two notable early case studies of children growing up with more than two languages that were widely cited in the early twentieth century: Milivoj Pavlovitch's (1920) study of a Serbian-French child and Jules Ronjat's (1913) study of a German-French bilingual child. Both men studied the language acquisition of their own sons, as the children learned the languages spoken by their parents. Pavlovitch, who was a Serbian native speaker, lived in France. His son Douchan was born on 13 October 1917 and was described as a healthy and fully developed newborn. Pavlovitch's main aim was to document the language acquisition of a bilingual child, and his book provides a detailed record of Douchan's first articulated utterances, repeated syllables, words, and sentences. Douchan learned Serbian first and Pavlovitch noticed that around the age of 14 months, when his son started to acquire French, the acquisition of his Serbian language skills slowed down. The later introduction of French also meant that it took several months before Douchan's Serbian and French vocabulary were equally well developed, and until he was two years old, Douchan continued to occasionally address people in the wrong language. From then on, however, he appeared able to distinguish between the two fairly reliably. There are other, more recent case studies of bilingual children who acquired languages sequentially and who acquired their second language rapidly and with relative ease (e.g. Fantini, 1985), supporting Pavlovitch's observations. The children in these studies were older than Douchan, although in some cases only slightly (e.g. Burling, 1959). However, as always, children develop at different speeds, and these case studies may merely reflect individual differences between children.

Ronjat's son, Louis, was born in 1908, as a healthy child. Ronjat was a linguist and specialised in the study of Occitan and not children's language development. However, as his wife was German, Ronjat found himself raising a bilingual child and seized the opportunity to research the topic of bilingualism further. Shortly before the birth of his son, his colleague Maurice Grammont (1902) had published his research on language acquisition in childhood. Ronjat consulted Grammont to seek his opinion on raising a bilingual child and some of this correspondence is reproduced in the book in which Ronjat discusses his son's language development, which was published when Louis was five years old. Grammont advised him to use what is now known as the "One Parent, One Language" method, which meant that Ronjat himself spoke French with Louis, while his wife spoke German with him. Louis did indeed become fluent in both languages and Ronjat's record of his son's development was frequently referred to in order to illustrate that the One Parent, One Language method could be used to raise a child bilingually. It should be mentioned, however, that other methods exist and, indeed, many parents find themselves switching between languages if their circumstances change (Pearson, 2008). This includes using different languages in different locations, also called "Time and Place", parents only speaking the minority language with their children, as the majority language is spoken in wider society (e.g. in schools), and parents mixing languages. However, the "One Parent, One Language" method appears to be successful in 74% of children, which suggests that it is a reliable and in most cases suitable approach for parents raising a bilingual child (de Houwer, 2009).

Neither Ronjat nor Pavlovitch noticed anything unusual about their children's non-linguistic development, which suggests that bilingualism had no effect on these cognitive functions. Similar to Dawes' (1902) and Williams' (1915) reports on the Belgian bilingual school system, the practical advantages of being bilingual were considered first and foremost, emphasising that the ability to speak several languages was considered desirable. However, the tides turned when researchers started to look into the effects of bilingualism on intelligence.

References

Barke, E. M., & Williams, D. P. (1938). A further study of the comparative intelligence of children in certain bilingual and monoglot schools in South Wales. *British Journal of Educational Psychology, 8*(1), 63–77.

Blommaert, J. (2011). The long language-ideological debate in Belgium. *Journal of Multicultural Discourses, 6*(3), 241–256.

Burling, R. (1959). Language development of a Garo and English speaking child. *Word, 15*(1), 45–68.

Caron, J. F. (2014). 2 Québécois and Walloon identities. Minority Nations in Multinational Federations: A comparative study of Quebec and Wallonia, 32.

Clark, G., & Henneberg, M. (2017). Ardipithecus ramidus and the evolution of language and singing: An early origin for hominin vocal capability. *Homo, 68*(2), 101–121.

Clayton, A. (2014). *The wars of French decolonization*. Routledge.

Dawes, T. R. (1902). *Bilingual teaching in Belgian schools, being the report on a visit to Belgian schools as Gilchrist travelling student*. Presented to the Court of the University of Wales.

De Houwer, A. (2009). Research methods in BFLA. In A. de Houwer (Ed.), *Bilingual first language acquisition*. Multilingual Matters.
Deprez, K. (1999). Flemish Dutch is the language of the Flemings. *Belgian Journal of Linguistics, 13*(1), 13–52.
Deschouwer, K. (2006). And the peace goes on? Consociational democracy and Belgian politics in the twenty-first century. *West European Politics, 29*(5), 895–911.
Deutscher, G. (2000). *Syntactic change in Akkadian: The evolution of sentential complementation*. OUP Oxford.
Devuyst, L. (1983). Moral education in Belgium. *Journal of Moral Education, 12*(1), 51–55.
Fantini, A. E. (1985). *Language acquisition of a bilingual child: A sociolinguistic perspective (to age ten)*. Multilingual Matters.
Gell-Mann, M., & Ruhlen, M. (2011). The origin and evolution of word order. *Proceedings of the National Academy of Sciences of the United States of America, 108*(42), 17290–17295.
Grammont, M. (1902). Observations sur le langage des enfants. In *Mélanges linguistiques. Offerts à M. Antoine Meillet par ses élèves* (pp. 61–82). Klincksieck.
Gxilishe, D. S. (1900). *Oral proficiency in Xhosa as a second language* (Doctoral dissertation, Stellenbosch University).
Hensley, D. J. (2008). *The blurred boundaries of Belgianness: Walloon intellectuals, pride, and the development of regionalist rhetoric, 1884-1914* (Doctoral thesis). Retrieved from https://etda.libraries.psu.edu/files/final_submissions/1411
Kroon, M., Medvedeva, M., & Plank, B. (2018, August). When simple n-gram models outperform syntactic approaches: Discriminating between Dutch and Flemish. In *Proceedings of the Fifth Workshop on NLP for Similar Languages, Varieties and Dialects* (VarDial 2018) (pp. 244–253).
Mettewie, L., & Mensel, L. V. (2020). Understanding foreign language education and bilingual education in Belgium: A (surreal) piece of cake. *International Journal of Bilingual Education and Bilingualism*, 1–19.
Naccache, A. F. H. (2016, March). *A social dimension of language evolution* [Conference Presentation]. Evolang 11 At: New Orleans, LA, USA Volume. Retrieved from http://evolang.org/neworleans/pdf/EVOLANG_11_paper_30.pdf
Pavlovitch, M. (1920). Le langage enfantin: Acquisition du Serbe et du Francais par un enfant Serbe.
Pearson, B. Z. (2008). *Raising a bilingual child*. Living Language.
Perreault, C., & Mathew, S. (2012). Dating the origin of language using phonemic diversity. *PLoS One, 7*(4), e35289.
Redknap, C., Lewis, W. G., Williams, S. R., & Laugharne, J. (2006). Welsh-medium and bilingual education. *Education Transactions, Series B: General*.

Ronjat, J. (1913). *Le développement du langage observé chez un enfant bilingue.* H. Champion.

Saer, D. J., Smith, F., & Hughes, J. (1924). *The bilingual problem.* Aberystwyth: University College Wales.

Schreyer, C. (2002). A proto-human language: Fact or fiction. *The University of Western Ontario Journal of Anthropology, 10*(1).

Singh, A. I. (1984). Decolonization in India: The statement of 20 February 1947. *The International History Review, 6*(2), 191–209.

Tarn, W. W. (1902). Notes on Hellenism in Bactria and India. *The Journal of Hellenic Studies, 22,* 268–293.

Thureau-Dangin, F. (1911). Notes assyriologiques. *Revue d'Assyriologie et d'archéologie orientale, 8*(1/2), 81–95.

van Gerwen, H., Bourguignon, M., & Nouws, B. (2017). Translating law in 19th-century Belgium: Criticisms of official translations of laws and decrees. *Tilburg Law Review, 22*(1–2), 99–137.

Williams, J. G. (1915). *Mother-tongue and other-tongue, or a study in bilingual teaching.* Jarvis and Foster.

Yoshioka, J. G. (1929). A study of bilingualism. *The Pedagogical Seminary and Journal of Genetic Psychology, 36*(3), 473–479.

CHAPTER 3

The Bilingual Problem

Abstract This chapter discusses early findings which suggested a negative effect of bilingualism on intelligence. In the light of this, some historic influences on early intelligence tests are considered and elaborated on. This is followed by a discussion of the first studies that identified methodological flaws in the design of studies that found a negative effect of bilingualism on intelligence.

Keywords Bilingualism · Intelligence test · Eugenics · Socio-economic status · Immigration

3.1 The Bilingual Problem

David John Saer was a head teacher at a boys' school in Aberystwyth, where he both taught and conducted research. In 1923, Sacr published a study in which he compared the performance of Welsh-English bilingual children on a range of tests to that of English monolingual children. While both groups of children performed equally well on a range of tasks, what concerned Saer was that the monolingual children performed far better on the intelligence test he had given them. Based on this difference in intelligence between the two groups of children, he concluded that the ability to speak more than one language may negatively affect other, non-linguistic domains, such as intelligence. To assess whether these impairments would be temporary or more permanent, Saer also compared a group of Welsh-English bilingual and English monolingual university students. Again, the monolinguals performed much better on the intelligence test.

Saer was not alone in his endeavour to study the effects of bilingualism on education. Around the same time, his colleague, Frank Smith, also published a study in which he administered intelligence tests to bilingual and monolingual children in Wales (Smith, 1923). Rather optimistically, he stated that the study was conducted with "the hope of testing the assertion that bilinguists are mentally superior to monoglots" (*p*. 273). Smith tested two groups of children, based on school years. The younger group would have been around eight years old while the older group would have been about 11 years old. Smith found that the younger children performed equally well on the intelligence tests, regardless of whether they were monolingual or bilingual. For the older children, however, the pattern previously observed by Saer repeated itself: the monolingual children performed much better on the task than the Welsh-English bilingual children. Smith's interpretation of the findings was that perhaps monolingual children develop their language skills faster than bilingual children between eight and 11 years of age.

What is important to note in regard to these two studies is that they were the first to ever intentionally set out to address whether bilingualism affects non-linguistic cognition in general and intelligence in particular. In other studies, the number of languages spoken by participants, and related to this, their fluency in these languages, was at times considered as a confounding variable, but it was not the central point of the investigation. For example, Brown (1922) tested the intelligence of children who had immigrated to the United States from various European countries. The schools he worked with recorded unusually poor performance across the different school years, and Brown hypothesised that perhaps one nationality may be of particularly low intelligence, thereby lowering education attainment. He acknowledges that language impacted some children's performance in his study with the following statement:

> In all cases, however, in which there was any doubt as to a pupil's ability to understand English sufficiently well to pass a test, he was given the test in his native language. Not infrequently we found children who, although they

spoke the English language fairly well, tested from six to eighteen months higher when their native language was employed. (pp. 324–325).

However, following this statement, the different native languages spoken by the children in this study are neither addressed in the results section of his article nor in the discussion of his findings. Brown does not seem to contemplate whether bilingualism may have had an effect on intelligence but instead takes it for granted that children should be provided with test materials in a language that will allow researchers to assess a their intellect with the greatest degree of validity and reliability. Following the publication of Saer's (1923) and Smith's (1923) findings, this approach would, for the most part, vanish from the literature.

While much research was conducted in regard to using intelligence tests in the context of immigration around the 1920s, which will be discussed in more detail shortly, Saer's and Smith's studies were the first to focus on language, and more specifically bilingualism, in a non-immigrant sample of participants. The findings reported in both pieces of research were combined and elaborated on in a more extensive record of the data, presented in book format and authored by Saer, Smith, and their colleague John Hughes (Saer et al., 1924). The book was titled *The Bilingual Problem* and composed with the aim to present the findings to interested lay people and those who would have use for them in their profession, specifically teachers. This increased the reach of the presented information, leading it to have a greater impact on both educational practices and research.

While Saer et al. (1924) acknowledged that, should these findings receive further support in the form of future research, this would have serious implications for the Welsh education system, they also cautioned their reader against any rushed decisions. They both emphasised the exploratory nature of their own research and highlighted that, as intelligence tests were still relatively new, it was possible that their findings occurred due to the effects of unintended methodological flaws in the study design. One such factor to consider would be that both Saer (1923) and Smith (1923) used the "rural-urban factor" to control for differences in socio-economic status, which would later be criticised as being an unreliable and uninformative way to account for differences in socio-economic status (e.g. Arsenian, 1937; Hill, 1936). To avoid all doubt in regard to this matter, they stated the following within the first few pages of *The Bilingual Problem*:

> No claim is made, and none can yet be made, for finality, and our discussion centres round the question of what the Welsh child is at present receiving from his school education, and not round that of the theoretical merits or demerits of bilingualism. This distinction is of importance in view of recent controversy, and it may yet be shown that bilingualism is an educational asset. (pp. 9–10).

Despite these precautions, *The Bilingual Problem* triggered a wave of research that subsequently investigated whether bilingualism affected intelligence in the population of bilinguals that were of interest to them. In this context, it is important to mention that the first intelligence tests only became available at the beginning of the twentieth century, and researchers were not yet aware of many of the background factors that could affect performance on these tests, for example socio-economic status (Boake, 2002). Much of the research that took place around the same time as Saer's (1923) and Smith's (1923) focused on immigration and intelligence, and only considered bilingualism in passing. It is, however, crucial to understand what effect the socio-political situation of immigrants had on how findings linked to bilingualism were interpreted in the context of intelligence. Section 3.2. will explore this in more detail. Other research focused on psychometric tests, with varying degrees of awareness of how participants' backgrounds may affect performance on these tests. Among these were investigations to address if participants performed differently on verbal and non-verbal intelligence tests. Saer and Smith had both utilised verbal intelligence tests but further research soon suggested that this might have put their bilingual participants at a disadvantage.

3.2 Early Intelligence Tests

This interest in non-linguistic cognition developed with the emergence of the first standardised psychometric tests, particularly intelligence tests. Since their inception, psychometric tests have been used to measure a range of different psychological constructs, for example personality traits (e.g. Vernon et al., 2008), mental health (e.g. Norton, 2007), and intelligence (e.g. Carpenter et al., 1990). They provide a standardised measure and generally aim to be objective, although differences in culture and socio-economic background can affect how questions and tasks contained within psychometric measures are perceived. Importantly, psychometric tests are convenient. They are often faster to use than, for example, conducting a long interview with participants. With appropriate training, they can be used by different experimenters at the same time, and they provide an easy to interpret score at the end of the task, which can be compared across different populations. Their convenience is particularly tempting now that computers record the responses, calculate participants' scores, and present researchers with the overview of all participant data at the end. If experiments are conducted online, there is, at least in theory, no need for the experimenter and the participant to ever exchange a single word beyond the consent form and experimental task.

More importantly, though, psychometric tests were novel at the beginning of the twentieth century. In general, if a new tool is available to scientists, most of us are keen to try it and apply it to our own specialty. And as with all new inventions, we generally need to learn a little more about them and tweak them accordingly, to get them to work as intended. Intelligence tests were no exception to this rule. The first test, the Binet—Simon Intelligence Scale, was released in 1905 by Alfred Binet and Theodore Simon. The scales continued to be reworked for a few years, with the final version finally released in 1908 (Wolf, 1969a). The test was, in part, developed in response to mandatory schooling in France and aimed to identify children who would not be fit for mainstream schooling, either because of developmental delays or due to cognitive deficits. Children with low intelligence were grouped into "morons" (70–51 IQ points), "imbeciles" (50–26 IQ points), and "idiots" (25–0 IQ points). The original classifications associated with the Binet—Simon scale may now sound strange or even offensive but were common and appropriate word choices for the times. The assumption was that the "imbeciles" and "idiots" would require care and specialist support, while the "morons" would be able to acquire sufficient skills and knowledge to perform simple tasks in order to earn a living and live mostly independently (Wolf, 1969b). Shortly before the first release of the Binet-Simon scale, in 1904, Binet joined a ministerial commission based on his previous work, which advocated a "special school" for "slow" children (Nicolas et al., 2013). Since mandatory education had been introduced in the late nineteenth century, children with low intelligence found themselves neither welcome in mainstream schools nor in hospitals, and neither of these two institutions appeared to feel particularly responsible for them either. To address this problem, Binet and Simon developed the intelligence scale with the aim to accurately identify children with cognitive deficits and developmental delays and, if appropriate, to proactively offer them an appropriate educational environment in specialised schools. The ministerial commission tasked with this endeavour was largely successful, not least because of Binet's research on intelligence.

The Binet—Simon scale was translated into English in 1916, by Lewis Terman, a psychologist working at Stanford University (Terman & Merrill, 1937). The English version of the test was subsequently named the "Stanford—Binet Test" and has at points been credited with moving measures of intelligence outside the psychiatric realm into the psychological one (Nicolas et al., 2013). Terman also incorporated a change proposed by a German psychologist, William Stern, who suggested that chronological age and the score achieved on the Binet—Simon scale should be assessed based on a ratio of these two measures, which would become known as the "Intelligence Quotient" (IQ; Lamiell, 2003, p. 61).

Thus, psychologists had a new and standardised tool to test cognitive ability, and as with all new tools, people were keen to try it. The Stanford—Binet Test came into existence at a time when the United States was experiencing an increased influx of immigrants, partially as a result of the two world wars. It was also a time in which socio-political forces increasingly pushed for English to be used as everyone's primary language, with an ever-decreasing number of bilingual schools in the United States (Gonzalez, 1975; Hakuta, 1986). This meant that immigrants were not only expected to rapidly learn English upon arrival but also, and perhaps more importantly, to raise their children in English. President Roosevelt could be heard saying that "if after say five years [the immigrant] has not learned English, he should be sent back to the land from whence he came" (as quoted in Gonzalez, 1975). It is safe to say that immigrants and refugees were not welcomed with open arms. At the same time, eugenics beliefs were on the rise and would continue to remain popular until well into the 1940s (Currell & Cogdell, 2006). Eugenics beliefs were linked to Nazi ideology, and while their popularity declined after World War II, they nevertheless had a strong influence on much of the psychological research conducted in the first half of the twentieth century (Kuhl, 2002).

In the light of the increase in immigration, the United States quickly became interested in developing a tool that would allow them to accept healthy and intelligent immigrants into the country and reject the ones they suspected would be less able to contribute to society. Intelligence tests seemed like a perfect tool to achieve this goal, and thus much of the early research with intelligence tests took place in the context of immigration. One issue with the Stanford—Binet test was that it largely relied on verbal ability, yet as we now know, bilinguals often perform worse on verbal intelligence tests than monolinguals (e.g. Darcy, 1953). This is not the case with non-verbal tests. In other words, if a bilingual and monolingual of equal intelligence both perform a verbal and non-verbal intelligence test, they would be expected to perform equally well on the non-verbal test but there is a very good chance that the monolingual would outperform their bilingual counterpart on the verbal intelligence test. This knowledge emerged early on, around the same time that Saer (1923) and Smith (1923) published their original research.

Initially, Pintner and Keller (1922), who were based in the United States, had tested bilingual and monolingual participants with the Stanford—Binet test and their own non-verbal test, the "Pintner Non-Language Test". For the Stanford—Binet test, their results follow the same pattern as Saer's and Smith's findings: bilinguals scored lower than the monolingual participants. For the Pintner Non-Language test, however, both groups performed equally well. A second study conducted by Pintner (1923) found similar results, using the Pintner Non-Language test and the National Intelligence Test used in the United States at the time, which relied mostly on verbal intelligence. As expected, the bilingual participants performed worse than the monolinguals on the test that primarily relied on verbal intelligence but both groups performed similarly on the non-verbal test. Upon arrival, immigrants often do not yet speak the language of their host country, which puts them at a disadvantage for verbal tasks. This was often the case for the "bilingual" immigrants, who may indeed not have been fully functionally bilingual after all. Colvin and Allen (1923) named this disadvantage "language handicap". In their research, they found that bilinguals and monolinguals achieved similar scores on simple arithmetic tasks, but for more complex tasks, the monolinguals they tested performed much better than the bilinguals. Colvin and Allen concluded that bilinguals' "language handicap" made it more difficult for them to understand and process the instructions for complex tasks, which in turn led them to perform worse. The bilinguals in their study also struggled with the verbal task they were given as part of the study, suggesting that they were indeed affected by a "language handicap". Finally, a contemporary literature review (Mead, 1926) on the use of intelligence tests with immigrant populations also arrived at the conclusion that verbal intelligence tests are not suited to adequately assess intelligence in bilinguals.

From a eugenics perspective, however, the interpretation of these findings appears to have been quite different. In many cases, researchers approached the differences observed for verbal and non-verbal tasks as an inability to adapt to different modes of testing, assuming that someone truly intelligent would be able to adapt to this difference without any ill effect. One extensive study with immigrants in the United States that followed this line of reasoning in its conclusions was conducted by Carl C. Brigham (1923). Brigham was born in the United States and was certainly at the beginning of his research career a firm believer in eugenics. He had served in World War I and, upon his return, joined Princeton University as a faculty member in 1920. For his research, he tested a large number of members of the military with mostly European immigration backgrounds. At the end of his book, he arrived at the following conclusion:

> The migrations of the Alpine and Mediterranean races have increased to such an extent in the last thirty or forty years that this blood now constitutes 70% or 75% of the total immigration. The representatives of the Alpine and Mediterranean races in our immigration are intellectually inferior to the representatives of the Nordic race which formerly made up about 50% of our immigration. In addition, we find that we are getting progressively lower and lower types from each nativity group or race. (p. 197).[1]

In his conclusion, Brigham highlights that the Caucasian Americans he tested performed best on the administered tests. He warned against the blending of races, as this superior intelligence of white Americans should be preserved, and highlighted that "pure races" were already rare to find in some countries, such as Switzerland and Belgium.

In Brigham's defence it should be said that seven years after publication of this work, he acknowledged that the interpretation of his findings was biased and led by prejudice (Brigham, 1930). He also acknowledges that many of the groups who scored low on the administered test were likely to speak English as a second language, which may have caused them to have been at a disadvantage. However, it looks as if this was too little, too

---

[1] In this context, "Alpine" refers to Eastern Europe and "Mediterranean" to Southern Europe, although points within Brigham's work suggest that he also considers Turkey to be part of this group, though this is never fully acknowledged. With "Nordic", he refers to Northern Europe.

late. A year after Brigham published his original findings, Harry Laughlin, an American sociologist who also earned the title "Superintendent of the Eugenics Record Office", was heavily involved in the congressional debate leading up to the passing of the 1924 Anti-Immigration Act (Bird & Allen, 1981). Laughlin used Brigham's findings to push for a stricter regulation of migration (Hawes, 1968; Herrnstein, 1995; Snyderman & Herrnstein, 1983). When the Act was passed, it primarily limited immigration from countries that had previously been associated with poorer cognitive performance on Brigham's tests and aimed to preserve the homogeneity of US citizens (Fairchild, 1924; Rausa, 2012; US Department of State, Office of the Historian, 2016). Thus, the development of intelligence tests was strongly influenced by eugenics beliefs and an interest in optimising immigration. While not necessarily the intention, this affected bilinguals negatively, as they were more likely to be immigrants than monolinguals.

Besides the effect of eugenics beliefs on how findings were interpreted, it was also unclear which, if any, background factors might act as confounding variables and to what extent these should be taken into account. Kimball Young completed his PhD in Psychology at Stanford University, from which he graduated in 1921. In the following year, his doctoral thesis titled "Intelligence tests of certain immigrant groups" was published. Young tested children, using verbal and non-verbal intelligence tests, and found that those with an immigration background were strongly affected by a language handicap. For context, the children he tested had arrived relatively recently in the United States and were still in the process of acquiring English. Young had hypothesised that children with "sufficient mental capacity" (p. 63) would overcome the language handicap, while others would not. How precisely Young defined sufficient mental capacity was not elaborated on, which means that it is not clear what specific cognitive processes he was referring to. The tests used in Young's research were the mostly verbal Army Alpha intelligence test and the non-verbal Army Beta test. The tests were originally developed to evaluate the potential of new recruits into the US military (Yoakum & Yerkes, 1920). The Alpha was given first to all literate recruits, while the Beta was given to recruits who failed the Alpha, who were illiterate, or who did not speak English. Scores were grouped into eight categories, with those achieving A, the highest category, marked as having potential to become higher officers. At the other end of the scale, the lowest two categories, D- and E, generally indicated that the men's mental age was below 10 years, although D- recruits were still considered to be suitable for regular service.

Young adapted the test for use with 12-year-old children and found that the Army Alpha was a better predictor of the children's educational achievements, and that bilingual children performed worse on it than monolingual children. This could of course be because a child's ability to express themselves will affect marks received for written assignments at school and their ability to understand complex arithmetic tasks will affect their performance in subjects such as maths. However, Young did not consider or acknowledge either of these potential explanations. Instead, he concluded that verbal intelligence tests should be used whenever possible, as these were better able to reflect real-life difference between people. Shortly after the publication of Young's doctoral research, Pintner (1923) discussed how verbal ability would affect children's educational performance, highlighting the difference in how researchers approached this topic. Young did acknowledge that there were substantial differences between the American parents and the parents of children who had recently immigrated to the United States. While 42.6% of the latter were earning a living as unskilled labourers, only 7.8% of the former did. We do know now that socio-economic status affects performance on many cognitive tests, including intelligence tests, which could mean that differences between the children of immigrants and non-immigrants had been exacerbated by differences in their socio-economic background (Arsenian, 1937; Morton & Harper, 2007). While Young did address the finding that parents' occupation was related to how well the children did on the tests, he dismissed the notion that this may have led to differences between the two groups of children rather rapidly: "it is sufficient to note that a considerable difference does exist, and must be taken into account in interpreting the total results of the study" (p. 30).

This highlights two very different approaches to the early study of verbal and non-verbal intelligence tests. Both Pintner and Keller (1922; Pintner, 1923) and Colvin and Allen (1923) were keen to emphasise that non-verbal tests are more suitable to compare bilinguals and monolinguals. In contrast, researchers who were closer to the eugenics movement considered verbal tests more suitable and emphasised that someone truly intelligent would be able to overcome or compensate for their language handicap. It is worth noting that Pintner (1923) cautioned researchers to be careful in the interpretation of their findings, as intelligence tests were still relatively new tools. However, later research did mostly agree that non-verbal tests should be used if the aim of the study is to compare bilinguals and monolinguals (see, e.g., Darcy, 1953).

An important point to note in regard to this early research is that many studies grouped participants into "immigrant" and "non-immigrant" based on names alone (Hakuta, & Feldman Mostafapour, 1996; Hakuta, 1986). In the context of the United States, if someone had a Spanish-sounding last name, they would be considered to be an immigrant, while someone with an American-sounding name would be assumed to have no immigrant background. The criteria that were applied to classify names in this way were rarely reported, but according to Hakuta and Feldman Mostafapour (1996), the majority of studies that used this method reported findings that suggest the monolingual group performed better on intelligence tests. There are two obvious issues with this approach. First, we cannot discern if someone has an immigrant background based on name alone. Second, even if we could accurately identify immigrants and non-immigrants, immigration status alone cannot tell us if someone is monolingual or not. At best, this method was unreliable but used in good faith. At worst, it suggests researchers may have been influenced by their own political agenda. Neither of these scenarios, or those that lie between the two extremes, are encouraging.

3.3 Late 1920s and 1930s

Besides the wave in research interested in the intelligence of immigrants, which led to an increasing body of research also investigating bilingualism, it took until the late 1920s for another study to be published that focused purely on the potential effects of bilingualism on intelligence. Joseph G. Yoshioka (1929) tested two groups of Japanese-English bilingual children living in California. The first group consisted of 17 children aged nine to 11 years; the second group included 21 children, aged from 12 to 15 years. Each child received a Japanese version and an English version of the National Intelligence Test as used in the United States at the time, and subsets of children also completed the Stanford–Binet test. For the children who also completed the Stanford–Binet test, Yoshioka compared their performance on the test to their performance on the National Intelligence Test in both languages. He found that, considering the IQ scores the children had obtained on the Simon–Binet test, they performed much lower on the National Intelligence Test than would be expected of them.

The normed National Intelligence Test scores obtained for children in the United States and Japan were based on research with monolingual children, and Yoshioka suggested that perhaps being bilingual may have limited the children's understanding of the test in either language. This would also explain why the older group of children, who had acquired higher levels of proficiency in both languages, deviated less from the normed test results than the younger group of children. The children in Yoshioka's sample performed within the expected range of intelligence scores, which suggests that being bilingual had no negative effect on the children's development. Additionally, Yoshioka was able to provide a plausible alterative explanation for unexpected observations. For example, the children fell below the American norms for the National Intelligence Test but the negative difference was even more extreme when the children were compared to the Japanese norms. However, as Yoshioka pointed out, the Japanese data used to establish norms for the test was collected in a particularly well-situated and educated area of Tokyo, which was likely to have led to inflated norms for the Japanese sample of children. This could easily explain the difference between Yoshioka's Californian children and the children tested in Japan. Of course, Pintner (1923) had also already discussed how the National Intelligence Test puts bilinguals at a disadvantage. In the penultimate paragraph of his article, Yoshioka addresses several of such matters before he presents his final verdict, which appears disconnected from the line of reasoning presented up to this point:

> It is suggested that bilingualism in young children is a hardship and devoid of apparent advantage, because bilingualism appears to require a certain degree of mental maturation for its successful mastery. (p. 479).

It is not clear how Yoshioka arrived at this conclusion based on the evidence he discussed, especially considering the number of alternative explanations he provided.

Meanwhile in the United Kingdom, the 1931 Committee for Research in Education listed "Influence of Bilingualism upon Mental Processes" under "Research Planned or in Progress" for the University of Bangor, reflecting that the topic had garnered the interest of a number of researchers. Shortly after, Gwilym R. Jones (1933 as cited in Baker, 1988) published his master's thesis, for which he made an effort to account for various background factors that may have affected children's performance. For his research, he matched Welsh-English bilingual children and English monolingual children, both groups around 10 years old, on measures of non-verbal intelligence and socio-economic status, and found no differences between them in regard to their ability to speak English. This then made it clear that bilingualism on its own was unlikely to disadvantage a child in their language development.

Furthermore, in 1931, David J. Saer's daughter, Hywella Saer,[2] who was a lecturer in Education at the University of Wales at the time, published research in which she aimed to introduce a standardised method to define the degree to which someone is bilingual. The system she proposed relied on proficiency but not on bilingual usage. In other words, it did not factor in whether the bilingual in question used both languages equally frequently (i.e. balanced bilingual) or whether one language was used more frequently than the other (i.e. unbalanced bilingual). The proposed system would see bilinguals based on the ratio of their proficiency in different languages. If the bilingual was equally proficient in both languages, the ratio would equal 1, and shift according to the differences in proficiency, which would allow researchers to classify the type of bilingualism more clearly and make it easier to compare findings across studies. Sadly, the system appears to have never been used in wider research, but this makes Hywella Saer the first person to address that bilingualism and monolingualism exist on a sliding spectrum and suggest a tool to reflect this.

This spectrum of bilingualism is also apparent in the sample of participants tested by Harry S. Hill (1936) for his research with Italian-English bilinguals only a couple of years later. The bilingual children in his study spoke Italian at home and had had limited exposure to English, for example at school, which led them to become less proficient in their use of English. The monolingual children in his study also came from Italian families but their parents spoke English at home. According to Hill, the monolingual children were exposed to the Italian language in their community but did not understand or speak Italian. He described them as having a good degree of English proficiency at school age. Among other factors, children were matched on socio-economic status, birth order, mental and chronological age, and gender. Italian-speaking children also completed a test of Italian language comprehension, to ensure that they were fully fluent in the language. Hill administered a range of intelligence tests and found that the two groups did not differ on measures obtained

[2] For anyone interested in the life of a headmaster and researcher in the 1920s, the Saer family photo album has been digitalised and can be found at the following address: bit.ly/2G2GIcB.

with the Pinter Non-Language Test and Haggerty Delta I Test (Haggerty, 1921). Nevertheless, Hill reported that the monolingual children performed better on tasks that required a moderate understanding of the task and a good command of English, while the bilingual children performed better on tasks that required more accuracy in addition to the appropriate use of English. On balance, however, Hill found no evidence of differences in intelligence between bilingual and monolingual children. Hill's research highlighted that if socio-economic factors were controlled for, bilingual and monolingual children performed equally well on intelligence tests.

Further support for this notion came from a landmark study conducted by Seth Arsenian (1937; Pintner & Arsenian, 1937), who matched high-proficient bilingual and low-proficient bilingual children on degree of bilingualism, language familiarity, age of acquisition, method of acquisition, and attitude towards the second language. The children in Arsenian's research consisted of 469 Jewish children in New York, who were familiar with Yiddish and English. The proficiency of the low-proficient children approached monolingual standards. The low-proficient and high-proficient children performed equally well on measures of intelligence, and Arsenian provided strong evidence for the effect of socio-economic status on performance on intelligence tests. He found that children with a higher socio-economic status generally performed better on intelligence tests than children from a lower socio-economic status, regardless of linguistic background. There was no effect of bilingualism on educational attainment, which strongly implied that the degree of bilingualism did not have a negative effect on children's mental capacities.

A year later, Ethel M. Marke and D. E. Perry Williams published another study of Welsh bilingual children. They included six schools in their study. Two of these schools were considered "bilingual" and were primarily attended by children who spoke Welsh at home and received Welsh instruction at the beginning of their schooling, with a gradual introduction of English into the curriculum. Another two "monolingual" schools were predominantly attended by children who spoke English at home and who received instruction in English. Finally, about half of the children who attended the remaining two schools spoke English at home while the other half spoke Welsh at home. The language of instruction for these "mixed" schools was Welsh, although they occasionally made use of English. The children completed a series of tests, including the 1932 Mental Survey Test of the Scottish Council, the Northumberland

Standardised Tests (General Intelligence version), the Non-Language Mental Tests, and Thorndike's Test of Word Knowledge. The three groups of children did not differ on the non-verbal intelligence test, that is, Pintner's Non-Language test, but the bilinguals performed much worse than the monolinguals on the verbal tasks. Barke and Williams interpreted this as strong evidence of a negative effect of bilingualism on intelligence.

To summarise, by the end of the 1930s, we already knew that socio-economic background affected data collected to study the effect of bilingualism on intelligence, as indicated by Arsenian (1937). Additionally, Hywella Saer (1931) was one of the first researchers to ask for a more reliable classification of what constitutes being bilingual. Pintner (1923; Pintner & Keller, 1922) had demonstrated that linguistic intelligence tests did not produce reliable results when bilinguals were tested, but some researchers considered this evidence that bilinguals were less adaptable to different test modalities and requirements of tests. This shows that in reality, they were less intelligent than their monolingual counterparts (Young, 1922). It would take several decades for the implications of these findings to be reflected in research design.

References

Arsenian, S. (1937). *Bilingualism and mental development.* College Press.

Bird, R. D., & Allen, G. (1981). The JHB archive report the papers of Harry Hamilton Laughlin, eugenicist. *Journal of the History of Biology, 14*(2), 339–353.

Boake, C. (2002). From the Binet–Simon to the Wechsler–Bellevue: Tracing the history of intelligence testing. *Journal of Clinical and Experimental Neuropsychology, 24*(3), 383–405.

Brigham, C. C. (1923). *A study of American intelligence.* Princeton University Press.

Brigham, C. C. (1930). Intelligence tests of immigrant groups. *Psychological Review, 37*(2), 158.

Brown, G. L. (1922). Intelligence as related to nationality. *The Journal of Educational Research, 5*(4), 324–327.

Carpenter, P. A., Just, M. A., & Shell, P. (1990). What one intelligence test measures: A theoretical account of the processing in the Raven Progressive Matrices Test. *Psychological Review, 97*(3), 404.

Colvin, S. S., & Allen, R. D. (1923). Mental tests and linguistic ability. *Journal of Educational Psychology, 14*(1), 1–20.

Currell, S., & Cogdell, C. (2006). *Popular eugenics: National efficiency and American mass culture in the 1930s.* Ohio University Press.

Darcy, N. T. (1953). A review of the literature on the effects of bilingualism upon the measurement of intelligence. *The Pedagogical Seminary and Journal of Genetic Psychology, 82*(1), 21–57.
Downey, M. T. (1961). *Carl Campbell Brigham: Scientist and educator.* Educational Testing Service.
Fairchild, H. P. (1924). The immigration law of 1924. *The Quarterly Journal of Economics, 38*(4), 653–665.
Gonzalez, J. M. (1975). Coming of age in bilingual/bicultural education: A historical perspective. *Inequality in Education, 19,* 5–17.
Haggerty, M. E. (1921). *Haggerty intelligence examination: Manual of directions for delta 1 and delta 2.* World Book Co.
Hakuta, K. (1986). *Mirror of language: The debate on bilingualism.* Basic Books.
Hakuta, K., & Feldman Mostafapour, E. (1996). Perspectives from the history and politics of bilingualism and bilingual education in the United States. In I. Parasnis (ed.), (pp. 38–50). New York and Cambridge: Cambridge University Press.
Hawes, J. M. (1968). Social scientists and the immigration restriction: High-lights of a debate, 1890-1924. *Journal of World History, 11*(1), 467.
Herrnstein, R. J. (1995). In R. Jacoby & N. Glauberman (Eds.), *The bell curve debate: History, documents, opinions.* Times Books.
Hill, H. S. (1936). The effect of bilingualism on the measured intelligence of elementary school children of Italian parentage. *The Journal of Experimental Education, 5*(1), 75–78.
Jones, G. R. (1933). Tests for the examination of the effect of bilingualism on 'intelligence' (Unpublished master thesis), cited in Baker, C. (1988). *Key issues in bilingualism and bilingual education.* Multilingual Matters.
Kuhl, S. (2002). *The Nazi connection: Eugenics, American racism, and German national socialism.* Oxford University Press.
Lamiell, J. T. (2003). *Beyond individual and group differences: Human individuality, scientific psychology, and William Stern's critical personalism.* Sage.
Mead, M. (1926). The methodology of racial testing: Its significance for sociology. *American Journal of Sociology, 31*(5), 657–667.
Morton, J. B., & Harper, S. N. (2007). What did Simon say? Revisiting the bilingual advantage. *Developmental Science, 10*(6), 719–726.
Nicolas, S., Andrieu, B., Croizet, J. C., Sanitioso, R. B., & Burman, J. T. (2013). Sick? Or slow? On the origins of intelligence as a psychological object. *Intelligence, 41*(5), 699–711.
Norton, P. J. (2007). Depression Anxiety and Stress Scales (DASS-21): Psychometric analysis across four racial groups. *Anxiety, Stress, and Coping, 20*(3), 253–265.
Pintner, R. (1923). Comparison of American and foreign children on intelligence tests. *Journal of Educational Psychology, 14*(5), 292–295.

Pintner, R., & Arsenian, S. (1937). The relation of bilingualism to verbal intelligence and school adjustment. *The Journal of Educational Research, 31*(4), 255–263.
Pintner, R., & Keller, R. (1922). Intelligence tests of foreign children. *Journal of Educational Psychology, 13*(4), 214–222.
Rausa, B. (2012). Immigration Act of 1924 (U.S.). In S. Loue & M. Sajatovic (Eds.), *Encyclopedia of immigrant health*. Springer.
Saer, D. J. (1923). The effect of bilingualism on intelligence. *British Journal of Psychology: General Section, 14*(1), 25–38.
Saer, H. (1931). An experimental inquiry into the education of bilingual peoples. In *British commonwealth education conference. Education in a changing commonwealth*.
Saer, D. J., Smith, F., & Hughes, J. (1924). *The bilingual problem*. University College Wales.
Smith, F. (1923). Bilingualism and mental development. *British Journal of Psychology: General Section, 13*(3), 271–282.
Snyderman, M., & Herrnstein, R. J. (1983). Intelligence tests and the Immigration Act of 1924. *American Psychologist, 38*(9), 986.
Terman, L. M., & Merrill, M. A. (1937). *Measuring intelligence: A guide to the administration of the new revised Stanford-Binet tests of intelligence*. Houghton Mifflin.
US Department of State, Office of the Historian. (2016). *The Immigration Act of 1924 (The Johnson-Reed Act)*.
Vernon, P. A., Villani, V. C., Vickers, L. C., & Harris, J. A. (2008). A behavioral genetic investigation of the Dark Triad and the Big 5. *Personality and individual Differences, 44*(2), 445–452.
Wolf, T. H. (1969a). The emergence of Binet's conception and measurement of intelligence: A case history of the creative process. *Journal of the History of the Behavioral Sciences, 5*(2), 113–134.
Wolf, T. H. (1969b). The emergence of Binet's conception and measurement of intelligence: A case history of the creative process. Part II. *Journal of the History of the Behavioral Sciences, 5*(3), 207–237.
Yoakum, C. S., & Yerkes, R. M. (1920). *Army mental tests*. H. Holt.
Yoshioka, J. G. (1929). A study of bilingualism. *The Pedagogical Seminary and Journal of Genetic Psychology, 36*(3), 473–479.
Young, K. (1922). Intelligence tests of certain immigrant groups. *The Scientific Monthly*, 417–434.

Chapter 4

Mid-Twentieth Century: Bilingualism and Intelligence

Abstract Research in the 1930s suggested that original findings that linked bilingualism to low intelligence occurred because background factors were not controlled for, estimates of bilingualism were unreliable, and the tests used to assess intelligence were not suitable to compare bilinguals and monolinguals. However, it took several decades before these findings translated into changes in research design and the theory that bilingualism causes "mental retardation" was dismissed.

Keywords Bilingualism · Intelligence · Socio-economic status · Education

Following the initial surge in research on bilingualism, the interest understandably calmed down during World War II, before picking up again in the 1950s and 1960s. The following will provide an overview of the significant findings reported during the mid- twentieth century.

4.1 1940s

The 1940s started with a study that compared Irish-English bilinguals to English monolingual children on measures of verbal intelligence (Stark, 1940). The children's ages ranged from 10 to 12 years and the bilingual children outperformed their monolingual peers. Additionally, Ojemann et al. (1941) reviewed factors that may affect mental development in children and arrived at the conclusion that the work they had reviewed did not suggest that bilingualism either benefits or harms intelligence. A similar review conducted by Tireman (1941) arrived at the same conclusion. In a particularly short article, Marshall and Phillips (1942) compared college grades of bilingual and monolingual students in the United States, and found that both groups of students performed equally well across all subjects. Not much later, Spoerl (1944) matched a group of bilingual and monolingual college students in the United States into pairs, based on the Henmon-Nelson Tests of Mental Ability, which are indicative of general mental ability. The two groups did not differ significantly in regard to their performance on the administered verbal intelligence test. However, the bilinguals' mastery of the English language was consistently higher than that of the monolinguals, and the bilingual students outperformed the monolinguals in terms of academic achievements. Thus, Tilden-Spoerl's findings suggested that bilingualism either had no effect on mental ability or a positive one—but it was unlikely to do any harm, at least in a college student population.

In the mid-1940s, Natalie T. Darcy had finished her PhD at Fordham University, for which she researched the intelligence of bilingual and monolingual children. Over the coming years, she would go on to publish a series of literature reviews and experimental work on the topic, the first of which was released in 1946. Darcy (1946) conducted a study with bilingual and monolingual children in New York. The children were matched on age, gender, and socio-economic background, and Darcy used both a verbal and a non-verbal intelligence test. The pattern of results for this study shows a fairly balanced picture: monolingual children performed better on the verbal intelligence test and bilingual children performed better on the non-linguistic test. However, Darcy focuses strongly on the monolinguals' superior performance on the verbal test and emphasises that the bilingual group appears to have been affected by a language handicap.

4.2   1950s

While studies early on found evidence that suggested that socio-economic status should be controlled for in research with bilinguals (e.g. Arsenian, 1937; Hill, 1936), researchers persisted with the use of the rural-urban factor and similar techniques for several decades (e.g. Carlson & Henderson, 1950). To assess the Welsh situation further, Jones and Stewart (1951) recruited several hundred Welsh-English bilingual children and English monolingual children to test their intelligence. The children were around 10–12 years old and the bilingual children were recruited from areas in which the community spoke predominantly Welsh, while the English children lived in an area in which the majority spoke English. The two groups were not matched beyond age, thus ignoring contemporary advice to match participants for socio-economic status. Jones and Stewart found that monolinguals performed better than the bilingual children on both the verbal and non-verbal intelligence tests. When they matched the two groups based on their performance on the non-verbal test, the difference between bilinguals and monolinguals was reduced but remained statistically significant. In a replication of the study, Jones (1953) later found that the bilingual and monolingual children in the new sample performed equally well on the non-verbal intelligence test but monolinguals performed better than bilinguals on the verbal intelligence test. However, based on the bilingual children's performance on the verbal test, as compared to the non-verbal test, Jones concluded that there was no evidence that the bilingual group was affected by a language handicap.

In terms of cultural assimilation, Johnson (1953) hypothesised that individuals who were better assimilated to their host culture would be better at speaking the local language. He tested 30 children, nine to 12 years old, who spoke both Spanish and English to varying degrees. As the study took place in the United States, Johnson suggested that better cultural assimilation would be marked by better English language skills and possibly poorer Spanish language skills. As part of his study, Johnson administered two intelligence tests, one that combined verbal and non-verbal measures and one that relied primarily on non-verbal measures. He found that a greater degree of bilingualism was associated with a negative outcome for the former and a positive result for the latter. Johnson did acknowledge that an intelligence test cannot be suitable to test for cultural assimilation, but nevertheless viewed these findings as tentative support for his hypothesis that better assimilated bilinguals perform better on intelligence tests. Importantly, Johnson's study also highlights the first tentative steps towards a more elaborate assessment of bilingualism in studies interested in its effect on non-linguistic domains. While Saer (1931) had already pointed out that bilingualism exists on a spectrum and should be treated as such, participants generally remained grouped into bilinguals and monolinguals, with only relatively basic tests administered to the bilinguals, to ensure that they truly spoke both languages. Johnson, however, used the "Hoffman Bilingual Schedule", which asks participants to indicate with what frequency they used each language in different situations (e.g. work, school) and with different people (e.g. parents, teachers; Hoffman, 1934). For example, Lewis (1959) used the same test to assess bilingualism when he studied differences in educational attainment between Welsh schools in bilingual areas and English-dominant areas. He found that educational attainment generally increased with pupils' intelligence but also that bilingual schools performed generally worse than English-monolingual schools. While Lewis explicitly acknowledged that the linguistic environment of a school may affect educational achievements, he did not provide a detailed discussion of potential reasons for this.

Both Kittell (1959) and Keston and Jimenez (1954) reported evidence that bilingual and monolingual children perform equally well on intelligence tests. Kittel tested children who were around seven to nine years old, using both a verbal and a non-verbal test. The bilingual children performed better than the monolinguals on the non-verbal test and the monolingual children outperformed the bilinguals on the verbal test. Kittell also accounted for socio-economic background, and his analysis suggests that one of the driving forces behind these differences was that monolingual middle-class children performed better than bilingual children who were in the middle or lower occupational class. Kittel emphasised that these socio-economic differences could act as confounding variables. As much of the previous research that investigated intelligence and bilingualism required bilingual children to complete intelligence tests in their second language, Keston and Jimenez had hypothesised that bilinguals would achieve lower scores because they completed these tests in a language they were less competent in. To investigate this hypothesis further, they tested English-Spanish bilingual children with two versions of the Stanford—Binet scale, one in English and one in Spanish. As the children's dominant language was English, they would thus be expected to perform better on the English version of the test, which is precisely what Keston and Jimenez found. It should be noted, however, that it only became apparent during the experiment that English was the children's dominant language. At the outset of the study, Keston and Jimenez expected children to have Spanish as their dominant language and thus expected that they would perform better when given the Spanish test.

While their results highlight that children's performance on intelligence tests depends at least partly on the language used for the test, they also emphasise the need to formally assess what a child's dominant language is first.

The first of two literature reviews by Natalie T. Darcy published in the 1950s became available in 1953. For the purposes of the review, she grouped studies into those that linked bilingualism to positive outcomes, those that linked bilingualism to a negative outcome, and those that found no effect of bilingualism on intelligence. In this context, however, it is worth noting that in the cases of some of the studies she reviewed, the bilingualism was only assumed and not assessed, which means the "bilinguals" may not have been bilingual. For example, she reviewed Davies and Hughes (1927), who compared the intelligence of Jewish children to the intelligence of non-Jewish children in London. They found that the Jewish children were, on average, a year ahead of the non-Jewish children in terms of intelligence. Darcy assumed that the Jewish children in the sample were bilingual, as their school was described as following Jewish traditions and customs. While this might well imply that the Jewish children spoke Yiddish or Hebrew in addition to English, it by no means implies that the non-Jewish group was monolingual. Darcy concluded her literature review by emphasising that verbal tests do appear to put bilingual participants at a disadvantage, while non-verbal intelligence tests do not. However, she stops short of recommending the use of one over the other.

4.3   1960s

At the beginning of the 1960s, Jones (1960) published a re-analysis of data he had first presented in the year prior (Jones, 1959). Jones had tested a group of Welsh-English bilinguals, who, similar to his previous studies (Jones, 1933, 1953; Jones & Stewart, 1951), performed worse on the administered intelligence test than their monolingual counterparts. When Jones (1960) re-analysed the data, he recorded socio-economic status for each of the children that was included in the study, based on their parents' occupational background. The key findings of this analysis were that, first, the monolingual children had, on average, a higher socio-economic status than the bilingual children, and second, when children from both groups were matched on socio-economic status, they performed equally well. This provided further support for the notion that socio-economic status, rather than bilingualism, may have been a driving factor behind the differences in intelligence observed for bilinguals and monolinguals, as had previously also been suggested by Arsenian (1937), Hill (1936), and Kittell (1959). Importantly, in his earlier studies, Jones repeatedly stated that children were recruited from schools in the same or neighbouring local district, which he assumed would prevent interference from environmental differences between children. In these studies, bilinguals consistently performed much worse on intelligence tests—verbal and non-verbal—than their monolingual counterparts. The findings reported by Jones (1960) strongly suggested that these earlier studies were affected by socio-economic status as an uncontrolled confounding variable, which implies that the low performance of bilinguals in these studies may in fact have been caused by differences in socio-economic status and not bilingualism. The study reported by Kittell (1959) provided another example of the effect of socio-economic background on performance on intelligence tests. He tested bilingual and monolingual children in third and fifth grade. When socio-economic status was not accounted for as part of the analysis, monolinguals outperformed bilinguals on all of the administrated measures of intelligence. If socio-economic status was accounted for, however, the bilingual fifth grade pupils performed better than their monolingual peers across all tests, and the third grade bilinguals and monolinguals matched in performance for most of the measures and social classes. The exception was the middle-class third grade children, for whom bilinguals performed worse than monolinguals on measures of intelligence. While this is another set of findings that highlights the importance of controlling for socio-economic status, surprisingly Kittell did not address this aspect of his findings in any depth. Instead, the discussion of his results focused on the findings that indicated a negative effect of bilingualism on intelligence when socio-economic status was not accounted for.

In the same year, Soffietti (1960) suggested that culture may affect whether bilingualism would lead to what had by then been coined "mental retardation". According to his definition, someone would be classed as unicultural if they only practised one culture and bicultural if they practised more than on culture. Soffieti reasoned that unicultural bilinguals would be less affected by mental retardation. He did not provide any data to support this claim but his logic followed a similar line of argumentation to that of Jones (1953), who suggested that successful cultural assimilation may lead to a reduction in negative effects of bilingualism on intelligence. The data presented by Jones generally supports this claim, but as discussed in the previous paragraph, his analysis did not account for background factors such as socio-economic status, which may well mean that the supporting data he presented cannot be fully relied on.

The hypothesis that bilingual children are able to perform more demanding cognitive tasks if they are well practised in a language was further strengthened by MacLeod (1969). He studied Scottish Gaelic-English bilingual children, who were more fluent in English than Gaelic. MacLeod reasoned that if frequent use of a language correlated with better performance on cognitive tests, monolinguals would have an advantage, as they did not need to split the time during which they spoke between two different languages. Not only would the monolinguals have more practice in using the one language they spoke, but they would also apply their language in more diverse contexts. The bilingual children MacLeod studied used English at school but spoke Gaelic outside of school. However, according to MacLeod's observations, the bilingual children were expected "to be seen, not heard" at home, which meant their opportunities to use Gaelic were limited to some extent. While this observation was likely to have been affected by MacLeod's own subjective judgement to a greater or lesser degree, it led him to emphasise that the social situation of bilinguals and monolinguals should be thoroughly acknowledged and matched as far as possible when we try to compare these two groups.

Interestingly, Macmeeken (1939) conducted a survey of children's intelligence across Scotland. Historically, the Western Isles, Skye, and Lochalsh contained the greatest proportion of Scottish Gaels, followed by the cities of Glasgow and Edinburgh (Baker, 1988). Macmeekan found that only the "industrial belt", which is more likely to be referred to as the "central belt" these days, performed notably worse than other areas of Scotland. There was no noteworthy difference between regions known to be inhabited by a larger number of Gaels and other areas of the country, suggesting that Gaelic culture did not negatively affect children's intelligence.

With regard to language, the data painted a less clear picture. Macmeekan did not record children's language status, which means that it is not clear whether the children he tested were bilingual or English-speaking monolinguals. The 1872 Education Act of Scotland had actively discouraged the use of Gaelic in education. Leading up to the passing of the Act, Gaelic had become stigmatised as being backward and schools often created an environment that was hostile towards Gaelic speakers, for example by appointing English-speaking monolingual teachers and punishing students for speaking Gaelic in school (Dunbar, 1999). Considering these difficulties, many Gaelic parents opted to not pass on Gaelic as a language to their children, which means that these children grew up as English-speaking monolinguals. The opinion that Gaelic education would be harmful continued until well into the twentieth century, with the first tentative movements towards a re-establishment of Gaelic education only emerging in the 1950s (Baker, 1988).

MacLeod's (1969) research is worth highlighting for yet another reason. He was the first to suggest that switching between local dialects may be linguistically similar to switching between languages. While he did not follow up on these musings, researchers interested in the effects of bilingualism on executive functioning would later aim to assess the same question in relation to their research (e.g. Antoniou et al., 2016; Kirk et al., 2014).

The perhaps most influential findings reported during the 1960s were in MacNamara's (1966) research in Ireland and a series of research reports linked to the Lambert Project in Canada, which aimed to develop a suitable immersion education programme. One example of the latter is Peal and Lambert's (1962) account of the intelligence of monolingual children and French-English bilingual children. The children were matched on a range of different background measures, including socio-economic status and gender. The bilingual children performed better than their monolingual peers on both verbal and non-verbal intelligence tests. Peal and Lambert interpreted their findings as support for the extension of French-language immersion education in Canada. The St Lambert Project got its name from the St Lambert community in Montreal, Canada, in which it took place. According to Baker (1988), who summarised the background of the St Lambert Project, the Anglophones in the community became increasingly frustrated with their inability to speak French and considered the methods by which second languages were taught at fault. Traditionally, the local schools had taught French as a second language for about half an hour every day, although the teachers giving French lessons were not always able to speak French fluently either. Much of the lesson was focused on rote learning and by the time pupils left school, only very few of them were able to functionally speak French. Throughout history, it has not been unusual for changes to language teaching to be influenced and sometimes led by initiations by parents. What makes the St Lambert Project stand out as unique is that the parents consulted with researchers at the local university on how to change second language education for the better. The experts in this case were Wallace Lambert and Wilder Penfield, who were based at McGill University and would continue to work with the community for many years.

As a first step, a language immersion kindergarten was set up in 1965, spaces for which reportedly filled within minutes (Lambert & Tucker, 1972). At the beginning of children's enrollment into the kindergarten, all of the daily lessons were taught in French, with English gradually being added to lessons. Children were encouraged to speak French if they chose to but were equally welcome to use English. There were no formal language lessons and no punitive measures if children chose to speak "the wrong language". To ensure the immersion programme did not have any unintended negative effects, Lambert and Tucker (1972) compared the children to a sample of children in mainstream education. Both groups of children were matched for socio-economic status and intelligence. They completed tests of achievements in both English and French, a maths test, as well as tests of attitude. Lambert and Tucker found no negative effects of the French immersion programme. The children in the programme were able to use English just as well as the children in mainstream education and the two groups performed similarly on all tests, except French language ability. The immersion class was able to use French equally well as English, which was not the case for the children in mainstream education.

The first class enrolled in the St Lambert experiment consisted of only 26 children and, according to Baker (1988), by 1986 120,000 children were enrolled in immersion education in Canada, with the prediction that by the early 1990s 20% of children would be enrolled in immersion programmes. While I have been unable to determine whether this prediction in terms of percentages was indeed correct, in terms of absolute numbers, at least 261,447 children were enrolled in French-English immersion programmes in the school year 1997/1998. However, the data for Quebec, Alberta, and Nunavut is missing for that year, which suggests that the true number may have been much higher (Statistics Canada, n.d., Table 37-10-0009-01). Immersion education continues to increase in popularity even today, which suggests that the St Lambert Project initiated a long-term change in how language in Canada is taught (Turcotte, 2019).

Meanwhile, in Ireland, MacNamara tested 1084 children who were enrolled in Irish-language schools and who spoke English at home. He recorded different measures of intelligence in Irish and English, assessed socio-economic background, and ask children to complete both a simple and a complex arithmetic task. If differences in non-verbal intelligence and socio-economic status were not accounted for, the bilingual children were approximately 17 months behind their monolingual peers on written tests, a difference that was reduced to 11 months if these factors were accounted for. The bilingual and monolingual children studied by MacNamara did not differ markedly in regard to how they performed the simple arithmetic task, but monolingual children performed much better on the complex arithmetic task than the bilingual group of children. MacNamara found that the Irish skills of children who spoke English at home were very weak, even after six years of instructions, which likely affected how competently they could express themselves in writing and how well they understood the instructions for the more complex arithmetic task. MacNamara's findings were subsequently used to discourage bilingual education in Ireland, while the findings reported by Peal and Lambert were used to support an increase in French immersion programmes in Canada.

The contrasting findings of these two studies emphasise the importance of MacLeod's (1969) conclusion, that in order to compare different groups of bilinguals and monolinguals it is important to consider their social situation, not just on an individual level but more widely. Whether a child successfully acquires a second language at school depends on whether the circumstances are suitable to do so. For example, the more proficient a language teacher, the more likely it is that students become functionally proficient in the language too (Alderson et al., 1997). The teaching of the Irish language in Ireland has a complex history, and the following can only provide the briefest of overviews in terms of its recent history. The period of the Great Famine (1845–49) put the Irish at risk of starvation and led many of them to emigrate in the hope of finding a more secure life elsewhere. The poorer Irish communities were more likely to speak Irish, yet they were also more vulnerable to the consequences of the Great Famine (Ceallaigh & Dhonnabhain, 2017). In that sense, the number of Irish native speakers decreased during this period due to death by starvation and an increased number of people who hoped to find reliable employment abroad and left the country. Ceallaigh and Dhonnabhain (2017) also argue that the famine made differences in social class more apparent and thereby led to the Irish language being associated with poverty and a lack of power. During the later nineteenth century, English increasingly became the language of commerce, and by extension employment, making the ability to speak English an attractive quality, yet Irish was placed on the school curriculum as an optional subject in 1878. At the beginning of the twentieth century, Irish experienced somewhat of a revival and the number of schools that taught Irish increased from 100 to 2000 between 1900 and the advent of the Irish Free State in 1921. Programmes promoting the Irish language were put into place, and by the 1940s, Irish was used as the medium of instruction in 12% of all primary schools and 28% of all secondary schools (Buachalla, 1984). However, a number of parents and teachers objected to the Irish language policies put into place and would have strongly preferred if English was taught as a first language, with Irish only introduced once children had mastered English (Baker, 1988). Following organised movements of parents, the number of Irish-medium schools dropped from 420 in 1960 to 160 in 1979, emphasising the preference for English as the language of instruction in schools (Buachalla, 1984).

In Canada, English and French did not have equal status until the 1969 Canada Official Language Act was passed. French was spoken by about a quarter of Canadians, with most of the French-speaking population concentrated in the province of Quebec, where English was in turn a minority language (Baker, 1988). One defining difference between the St Lambert experiment and its associated research and MacNamara's (1966) research is who wanted to learn the minority language. Irish had become a minority language in Ireland during the nineteenth century but there was little social, political, or economic incentive to gain from learning Irish for someone who spoke English (Kallen, 1988). Meanwhile, French across Canada was considered a minority language, and even in Quebec French speakers were reported to be more likely to switch to English than to insist on French. Yet it was the English-speaking minority in Quebec that initiated the move towards French immersion schools (Baker, 1988). Thus, the English-speaking Canadians who enrolled their children in French immersion classes did so willingly and enthusiastically, while in Ireland most English-speaking students only reluctantly enrolled in Irish-medium education. Similarly, while there is a good body of evidence that teachers in Ireland were somewhat disgruntled with the requirement to teach in Irish, there appears to be no such record of the Canadian teachers involved with the French immersion programme (Baker, 1988). Lastly, the development of the French immersion programme was guided by consistent research that accompanied it, and is likely to have benefitted from the data that was collected over time, repeatedly showing that bilingual education did not disadvantage children.

References

Alderson, J. C., Clapham, C., & Steel, D. (1997). Metalinguistic knowledge, language aptitude and language proficiency. *Language Teaching Research, 1*(2), 93–121.

Antoniou, K., Grohmann, K. K., Kambanaros, M., & Katsos, N. (2016). The effect of childhood bilectalism and multilingualism on executive control. *Cognition, 149*, 18–30.

Arsenian, S. (1937). *Bilingualism and mental development*. College Press.

Baker, C. (1988). *Key issues in bilingualism and bilingual education*. Multilingual Matters.

Barnhart, W. R., Rivera, S., & Robinson, C. W. (2018). Effects of linguistic labels on visual attention in children and young adults. *Frontiers in Psychology, 9*, 358.

Buachalla, S. O. (1984). Educational policy and the role of the Irish language from 1831 to 1981. *European Journal of Education, 19*(1), 75.

Carlson, H. B., & Henderson, N. (1950). The intelligence of American children of Mexican parentage. *The Journal of Abnormal and Social Psychology, 45*(3), 544–551.

Ceallaigh, T. Ó., & Dhonnabhain, Á. N. (2017). Reawakening the Irish language through the Irish education system: Challenges and priorities. *International Electronic Journal of Elementary Education, 8*(2), 179–198.

Darcy, N. T. (1946). The effect of bilingualism upon the measurement of the intelligence of children of preschool age. *The Journal of Educational Psychology, 37*, 21–44.

Davies, M., & Hughes, A. G. (1927). An investigation into the comparative intelligence and attainments of Jewish and non-Jewish school children. *British Journal of Psychology, 18*(2), 134.

Dunbar, R. D. (1999). The conflict between national and minority. In D. Fottrell & B. Bowring (Eds.), *Minority and group rights in the new millennium*. Martinus Nijhoff Publishers.

Hill, H. S. (1936). The effect of bilingualism on the measured intelligence of elementary school children of Italian parentage. *The Journal of Experimental Education, 5*(1), 75–78.

Hoffman, N. N. H. (1934). *The measurement of bilingual background. Contributions to education*. Bureau of Publications, Teachers College, Columbia University.

Johnson, G. B. (1953). Bilingualism as measured by a reaction-time technique and the relationship between a language and a non-language intelligence quotient. *The Pedagogical Seminary and Journal of Genetic Psychology, 82*(1), 3–9.

Jones, G. R. (1933). *Tests for the examination of the effect of bilingualism on 'intelligence'* (Unpublished master thesis), cited in Baker, C. (1988). *Key issues in bilingualism and bilingual education*. Multilingual Matters.

Jones, G. R. (1953). The influence of reading ability in English on the intelligence test scores of Welsh-speaking children. *British Journal of Educational Psychology, 23*(2), 114–120.

Jones, G. R. (1959). *Bilingualism and intelligence*. University of Wales Press.

Jones, G. R. (1960). A critical study of bilingualism and non-verbal intelligence. *British Journal of Educational Psychology, 30*(1), 71–77.

Jones, G. R., & Stewart, W. A. C. (1951). Bilingualism and verbal intelligence. *British Journal of Statistical Psychology, 4*(1), 3–8.

Kallen, J. L. (1988). The English language in Ireland. *International Journal of the Sociology of Language, 1988*(70), 127–142.

Keston, M. J., & Jimenez, C. (1954). A study of the performance on English and Spanish editions of the Stanford-Binet intelligence test by Spanish-American children. *The Journal of Genetic Psychology, 85*(2), 263–269.

Kirk, N. W., Fiala, L., Scott-Brown, K. C., & Kempe, V. (2014). No evidence for reduced Simon cost in elderly bilinguals and bidialectals. *Journal of Cognitive Psychology, 26*(6), 640–648.

Kittell, J. E. (1959). Bilingualism and language—Non-language intelligence scores of third-grade children. *The Journal of Educational Research, 52*(7), 263–268.

Lambert, W. E., & Tucker, G. R. (1972). *Bilingual education of children: The St. Lambert experiment*. Newbury House.

Lewis, D. G. (1959). Differences in attainment between primary schools in mixed language areas: Their dependence on intelligence and linguistic background. *British Journal of Educational Psychology, 30*, 63–70.

MacLeod, F. (1969). *An experimental investigation into some problems of bilingualism* (Doctoral dissertation). Retrieved from http://digitool.abdn.ac.uk/R?func=search-advanced-go&find_code1=WSN&request1=AAIU602195

Macmeeken, A. M. (1939). *The intelligence of a representative group of Scottish children*. University of London Press.

MacNamara, J. T. (1966). *Bilingualism and primary education*. The University Press.

Marshall, M. V., & Phillips, R. H. (1942). The effect of bi-lingualism on college grades. *The Journal of Educational Research, 36*(2), 131–132.

Ojemann, R. H., Garrison, K. C., & Jensen, K. (1941). Mental development from birth to maturity. *Review of Educational Research, 11*(5), 502.

Peal, E., & Lambert, W. E. (1962). The relation of bilingualism to intelligence. *Psychological Monographs: General and Applied, 76*(27), 1–23.

Saer, H. (1931). An experimental inquiry into the education of bilingual peoples. In W. T. R. Rawson (Ed.), *Education in a changing commonwealth (Report of a British commonwealth education conference in London)* (pp. 116–121). New Education Fellowship.

Soffietti, J. P. (1960). Bilingualism and biculturalism. *The Modern Language Journal, 44*(6), 275–277.

Spoerl, D. T. (1944). The academic and verbal adjustment of college age bilingual students. *The Pedagogical Seminary and Journal of Genetic Psychology, 64*(1), 139–157.

Stark, W. A. (1940). The effect of bilingualism on general intelligence: An investigation carried out in certain Dublin primary schools. *British Journal of Educational Psychology, 10*, 78–79.

Statistics Canada. (n.d.). *Table 37-10-0009-01 Number of students in official languages programs, public elementary and secondary schools, by program type, grade and sex.* Retrieved from https://doi.org/10.25318/3710000901-eng

Tireman, L. S. (1941). Chapter VIII: Bilingual Children. *Review of Educational Research, 11*(3), 340–352.

Turcotte, M. (2019). *Results from the 2016 Census: English-French bilingualism among Canadian children and youth.* Statistics Canada: Statistique Canada.

CHAPTER 5

Late Twentieth Century: Meta-Linguistics

Abstract During the late twentieth century, the focus of bilingualism research shifted away from intelligence and moved towards meta-linguistics. Based on this meta-analytic research, the hypothesis emerged that bilinguals may experience wider, non-linguistic executive functioning advantages. By the end of the century, the term "Bilingual Advantage" had been coined.

Keywords Bilingualism • Executive functioning • Meta-linguistics • Attention • Replication crisis

5.1 1970s

During the middle of the twentieth century, meta-cognitive research, in other words research on how we think about thoughts, became an increasingly popular topic in psychology (Goldberg, 1963; Smith, 2001). This trend finally reached research interested in the effects of bilingualism in the 1970s, although it focused more on meta-linguistic than meta-cognitive processes, that is, the ability to reflect on the use of language. While the focus of the present work is investigations with an interest in how bilingualism affects non-linguist behaviour, the meta-linguistic work that was conducted between the 1970s and 1990s eventually led to the first tentative studies that investigated the effects of bilingualism on executive functioning. It also provided further insights into how the degree to which someone is bilingual or monolingual may affect the outcome of studies, and thus directly relates to more recent efforts to develop a reliable classification of bilinguals and monolinguals, which in turn would allow for a more reliable comparison of findings across different studies. For these reasons, the following will provide a brief overview of the key articles published on the topic of bilingualism and meta-linguistics.

One of the most relevant findings of this period is that bilinguals were found to be better able to manipulate and change linguistic labels and struggled less with any ambiguity of linguistic labels (Cummins & Mulcahy, 1978; Feldman & Shen, 1971; Ianco-Worrall, 1972). Linguistic labels are generally words that refer to categories (e.g. "humans" includes men, women, and children) or objects (e.g. "car", "book"). The labels used in research are often invented for the purposes of the study. For example, Barnhart et al. (2018) used a task that introduced children to pictures of creatures "from another planet" and taught them the creatures' names, for example "dax", a word that would have been unfamiliar and new to the children in the study. That bilinguals found it easier to manipulate linguistic labels may also be linked to greater cognitive flexibility, something that Bain (1978) reported evidence for. He compared the performance of children enrolled in either a language immersion programme or educational music programme on a range of verbal and non-verbal tasks. The children enrolled in the immersion programme performed better on the administered verbal tasks than the children enrolled in musical training, while the latter performed better on tasks that required inner awareness. Bain interpreted these findings as a sign that both musical training and language learning can lead to increased cognitive flexibility.

In general, it appeared that bilinguals had a much better awareness of how they themselves and others used language, as compared to their monolingual peers (Cummins, 1978). Additionally, Cummins (1979) proposed the "threshold hypothesis" of bilingualism. According to his theory, children who progressed past a certain point of proficiency, or in other words "crossed the threshold", would not experience any negative effects of bilingualism in their cognitive development. Indeed, Cummins suggested that children who crossed this threshold might greatly benefit from being bilingual. The threshold hypothesis would not stand the test of time, but it is worth mentioning as one of the first theories that explicitly suggested that second language proficiency might be linked to whether bilingualism has a positive or negative effect on cognition (Rolstad & MacSwan, 2014). Cummins' (1979) article also drew attention to the fact that bilinguals require two labels for every object they encounter. For a German-English bilingual, that might be "tisch" and "table" to describe a table. Not much later, Oren (1981) found that bilinguals find it easier to manipulate object labels, which could certainly be linked to their experience of managing the use of multiple different object labels for the same object in day-to-day life.

5.2 1980s

In a review of the early literature on bilingualism and intelligence, and the more recent meta-linguistic work, Diaz (1983) arrived at the conclusion that the lack of control for socio-economic background was likely to have acted as a confounding variable in the majority of research. Following his literature review, Diaz (1985) published a longitudinal study with two groups of Spanish-English bilingual children, aged five to seven years old, who were tested twice. Based on Cummins' (1979) threshold hypothesis, Diaz predicted that children who were less fluent in their second language would encounter greater difficulties with mental tasks. Contrary to this prediction, on the first occasion of testing, Diaz found that degree of bilingualism was positively correlated with performance on measures of intelligence and cognitive functioning. Within the range of proficiency of the high-proficient children, degree of bilingualism did not appear to be correlated with performance on the administered test battery. Although Diaz did report evidence that lower socio-economic status was linked to poorer performance on the battery of tests, by the time the children were tested for the second time, the low-proficient bilingual children with a lower socio-economic status matched the performance of the high-proficient bilinguals on almost all measures. Diaz interpreted these findings as bilingualism preventing some of the potentially negative impact of low socio-economic status on cognitive development.

Using a grammatical judgement task, Bialystok (1986) tested bilingual and monolingual children's grammatical judgement on sentences that required either high or low levels of linguistic control. Both groups of children performed equally well if the sentence required a low level of linguistic control, but bilinguals performed better than monolinguals on sentences that required a high level of linguistic control to perform the task. Based on this observation, Bialystok concluded that the bilingual children may have better cognitive control. Building on this work, Bialystok (1988) presented two studies. In the first, she tested monolingual children, fully bilingual children, and partially bilingual children. The partially bilingual children spoke English as their native language but were enrolled in a French immersion school. All three groups of children lived in Canada and they were between 6.5 and seven years old at the time of testing. This meta-linguistic study presented the children with three different tasks, one to investigate how the children she tested dealt with the arbitrariness of language, one to assess their knowledge of abstract words, and one that required children to correct grammatically incorrect sentences. For the task that assessed arbitrariness of language, Bialystok used the moon/sun problem originally proposed by Piaget (1929). The children were asked to imagine what would happen if we all suddenly decided to call the moon "sun" and the sun "moon". They then had to answer questions such as "What would you call the thing in the sky when you go to bed?" to determine whether the temporary agreement to switch labels for that object would carry over to the child's response. To give the correct answer on this task, children need to inhibit the label they would usually apply to the object in the question. They have to ignore the fact that, usually, that object is called moon and instead choose "sun" to answer the question correctly. The second part of this task required the children to place this new label within the changed circumstances, by indicating what the sky looks like when they go to bed (i.e. dark). The second task, used to assess the children's conceptual understanding of words, presented them with a list of words and asked them to, first, identify if the items on the list were real words and, second, to define the words. The third and final task was to identify and correct grammatical errors in spoken sentences.

Bialystok (1986) found that the fully bilingual groups of children performed significantly better on the task that required them to identify and correct grammatical errors in spoken sentences than the other two groups, but there was no difference between groups for the task that asked them to identify words as real words and define them. However, there was a difference between groups for the moon/sun task: the two bilingual groups performed the task much more successfully than the monolingual group, suggesting that they were better able to inhibit interference from the object label they would usually use. Notably, however, the groups of children did not differ on the second part of this task, which asked them to switch the labels for "dog" and "cat". Bialystok explained this by suggesting that children were more familiar with cats and dogs than with the moon and the sun, and thus found it more difficult to change the labels of the animals. At the descriptive level, the monolingual children found it easier to change labels for the dog/cat problem than for the sun/moon problem, while the opposite pattern emerged for the bilingual children. For the cat/dog version, the children were also asked what sound the animal in question would make; for example, while looking at the picture of a cat, they would be expected to call it a dog while also indicating that the sound it makes is "meow". Bialystok argued that, consequently, the task was not strictly meta-linguistic but incorporated aspects of concept-formation.

For the second study she reports, Bialystok (1988) removed the dog/cat problem and carried on only with the moon/sun version of the task. Her reason for this was that, as mentioned above, she assumed that the cat/dog version of the task was not a pure meta-linguistic task and that "[b]ecause all three groups responded in the same way to the cat/dog version, it is possible that the task is not equivalent to the sun/moon problem" (p. 564). She also excluded the syntax correction task but added a grammatical judgement task. The latter required children to judge sentences as either grammatically correct or semantically correct. The sentences the children were presented with were grammatically correct but semantically incorrect (e.g. "the chair ate an apple"), semantically correct but grammatically incorrect (e.g. "the ate boy an apple"), or incorrect both in terms of grammar and meaning (e.g. "the ate chair an apple"). The children, however, were only asked to identify grammatical errors, which can be difficult to do in the presence of interfering semantic information, which often receives more attention than grammar. These days, if you use social media you may well be familiar with the various memes along the lines of "After reading this sentence, you will realise that the the brain ignores the second 'the'", which illustrates that detecting grammatical errors can be a very difficult task.

For the children in Bialystok's (1988) study, this meant having to ignore the semantic knowledge they had and only focus on the grammatical information available to them. The children in this study were the same age as the children tested in the first study but consisted only of high-proficient and low-proficient English-Italian bilinguals living in Canada. Bialystok hypothesised that the grammatical judgement task condition (labelled "anamalous") measured cognitive control. This condition presented children with grammatically correct but semantically incorrect sentences which meant attention usually directed towards sentence meaning needed to be suppressed in order for the children to focus on grammar. The opposite condition, that is, grammatically incorrect but semantically correct (labelled "incorrect"), was considered as an indication of the children's analysis of knowledge. Meanwhile, she considered the moon/sun problem a measure of control of processing. The only one of these measures on which the low-proficient and high-proficient bilinguals differed was the analysis of knowledge. They did not differ on any of the other two measures thought to be indicative of cognitive control processes. Based on these findings, Bialystok concluded that different degrees of bilingualism conferred different degrees of meta-linguistic advantage, but she also discussed how bilingualism may lead to benefits of cognitive control in other domains:

> Other benefits may accrue from advanced levels of control of processing. If such processing is general to other cognitive domains and not restricted to linguistic processing, certain spatial problems may involve the same skill. […] If isolating and attending to formal constituents of larger meaningful constructs is a generalized ability, bilingual children may perform in a more field-independent way than monolingual children. (p. 566).

The 1988 study conducted by Bialystok stands out as the first in which she explicitly discusses a potential link between bilingualism and cognitive control in a wider sense, a hypothesis that would inform much of her research in the next decade.

5.3 1990s

Bialystok (1992) followed up on her hypothesis that bilingualism may be linked to cognitive control in a detailed review article in which she discussed theories of attention and bilinguals' enhanced performance on tasks that require meta-linguistic control. She hypothesised that the superior linguistic control she observed for bilinguals as compared to monolinguals might generalise to non-cognitive domains. The explanation she proposed was based on the fact that bilinguals have to constantly manage access to two different languages. If they view an object, they cannot simply default to one label to describe the object they look at but instead have to choose one of several options, based on their environment and context. A Dutch-English bilingual trying to ask for a pencil has to decide whether to ask for a "pencil" or "potlood" in order to communicate effectively. In this moment, "pencil" and "potlood" compete for representation and we can use our ability to deploy attention effectively to choose the right object label. Bialystok argued that this "practice" of using attention effectively happens any time a bilingual uses language and may thus make attentional processes more effective in bilinguals. Based on her meta-linguistic findings, she suggested that this attentional advantage may generalise to other, non-linguistic, domains of cognition. Bialystok presented a compelling case, but her theory was still new and remained to be tested.

Bialystok did present data in support of her theory later on in the 1990s. First, Bialystok and Majumder (1998) presented data collected with English monolingual children, low-proficient Bengali-English bilingual children, and high-proficient French-English bilingual children. The children were described as coming from middle-class households.[1] All three groups performed three problem-solving tasks, including two that contained misleading information (Block Design Task, Wechsler, 1974; Piaget's Water Level Task, Pascual-Leone, 1969) and one that did not include misleading information (Noelting Juice Task, Noelting, 1980a, 1980b). If bilinguals are really better at focusing their attention on relevant information while inhibiting distracting information, bilinguals should have performed better than monolinguals on the tasks with distracting information, but all groups would be expected to perform equally well on the problem-solving task without distracting information. For one of the tasks with distracting information (Block Design Task, Wechsler, 1974), the French-English bilinguals performed better than the other two groups. On the second task with distracting information (Piaget's Water Level Task, Pascual-Leone, 1969), differences between groups "approached significance", primarily driven by the French-English bilinguals performing better than the other two groups. There were no group differences for the third task without distracting information.

Throughout the twentieth century, the statistical analysis of psychological data became increasingly more advanced. For example, Baker

---

[1] The children are described as balanced (French-English) and unbalanced (Bengali-English) bilinguals in the original article by Bialystok and Majumder (1998). However, the tests they conducted suggest that the main difference between the bilingual groups was fluency. The French-English bilinguals were fluent in both languages, while the Bengali-English children were fluent in English but spoke Bengali to varying degrees. They were regularly exposed to Bengali, which further supports that the deciding factor between groups was language proficiency rather than language use.

(1988) praised MacNamara's (1966) use of regression analysis, which at the time was rarely used in psychological or linguistic research. By the 1970s and 1980s, statistical methods that provided a p-value to discern if differences between the categories of interest were significant were generally standard in quantitative psychological research. Significance testing in the social sciences in general, and psychology in particular, has been criticised since it started to be used more widely (e.g. Bakan, 1966; Carver, 1978; Cohen, 1990; Lykken, 1968). The issue with significance testing was, and remains, that researchers' understanding of p-values has room for improvement.

In the late 1980s, Oakes (1986) presented 68 academics in psychology with a choice of six definitions of the p-value. All of them were inaccurate, yet only 3% of the 68 academics who responded accurately identified all six definitions as wrong. Haller and Krauss (2002) repeated the experiment with different groups of psychologists: 30 psychologists who taught psychological research methods and statistics at university, 39 who were psychological researchers, and 44 who were psychology students. Of the psychologists who taught research methods, 20% correctly identified all responses as wrong; of the psychological researchers, 10.3% responded correctly and 0% of the students got it right. Textbooks are unlikely to come to the rescue of our statistical honour, either. Haller and Krauss looked at one example in more detail (Nunnally, 1975) and discovered eight different interpretations of statistical significance within the space of three pages. While it would be wonderful to think that things have changed since the 1970s, a more recent study found that only around 11% of textbooks for research methods in psychology correctly define the statistical significance as indicated by a significant p-value (Cassidy et al., 2019). In short, as psychologists we have not exactly covered ourselves in glory when it comes to understanding the statistics we use on a near-daily basis.

While the $p = 0.05$ threshold is widely accepted as the threshold for significance, phrases similar to "showed a trend towards significance", "marginally significant", or "approached significance" have been frequently used in the case of p-values that fell a little short of crossing this threshold. Increasingly, there are efforts to rectify this situation, for example by considering the size of the effect under investigation, as well as different statistical measures together instead of focusing on p-values more or less exclusively (e.g. Calin-Jageman & Cumming, 2019; Lakens, 2019). However, while these days phrases such as "approaching significance", as used by Bialystok and Majumder (1998) to describe the findings for one of the tasks with distracting information, would be viewed as poor practice by some, this was a common description of statistical findings for many decades. It is thus not surprising that Bialystok and Majumder interpreted the finding that the French-English bilinguals performed significantly better on one problem-solving task with distracting information and almost significantly better on the other one as support for the conclusion that bilinguals, at least high-proficient bilinguals, have better problem-solving skills than monolinguals.

Following on from problem-solving skills, Bialystok (1999) tested three-to-five-year-old children on a task-switching task, more specifically the Dimensional Change Card Task (DCCT; Zelazo & Frye, 1997). The task requires children to sort cards according to one dimension, for example colour of objects shown in the first round of trials and another dimension, for example. shape, in the second block of the experiment. Thus, if sorted correctly, a "green triangle" would be sorted into the "triangle pile" in the first phase but the "green pile" in the second phase. In the third phase, the experimenter asks children to discern whether they understood the rules. In general, children find it difficult to change from sorting by colour to sorting by shape or vice versa. Bialystok found that the bilingual children made significantly fewer errors than the monolingual children. She interpreted this finding as support for her hypothesis that the cognitive demands of bilingualism provide a "training effect" for attention: that generalised to non-linguistic cognitive domains. Notably, this article also saw a change in terminology as her theory developed. While she had previously discussed non-linguistic attention as a domain to which the benefits of bilingualism may generalise, she now used the term "executive functioning", which refers to both attention and working memory (Goldstein et al., 2014). With this study, the "Bilingual Advantage" was born.

References

Bain, B. (1978). The cognitive flexibility claim in the bilingual and music education research traditions. *Journal of Research in Music Education, 26*(2), 76–81.

Bakan, D. (1966). The test of significance in psychological research. *Psychological Bulletin, 66*(6), 423.

Baker, C. (1988). *Key issues in bilingualism and bilingual education* (Vol. 35).

Barnhart, W. R., Rivera, S., & Robinson, C. W. (2018). Different patterns of modality dominance across development. *Acta Psychologica, 182*, 154–165.

Bialystok, E. (1986). Factors in the growth of linguistic awareness. *Child Development, 57*, 498–510.

Bialystok, E. (1988). Levels of bilingualism and levels of linguistic awareness. *Developmental Psychology, 24*, 560–567.

Bialystok, E. (1992). Selective attention in cognitive processing: The bilingual edge. *Advances in Psychology, 83*, 501–513.

Bialystok, E. (1999). Cognitive complexity and attentional control in the bilingual mind. *Child Development, 70*, 636–644.

Bialystok, E., & Majumder, S. (1998). The relationship between bilingualism and the development of cognitive processes in problem solving. *Applied Psycholinguistics, 19*, 69–85.

Calin-Jageman, R. J., & Cumming, G. (2019). The new statistics for better science: Ask how much, how uncertain, and what else is known. *The American Statistician, 73*(sup1), 271–280.

Carver, R. (1978). The case against statistical significance testing. *Harvard Educational Review, 48*(3), 378–399.

Cassidy, S. A., Dimova, R., Giguère, B., Spence, J. R., & Stanley, D. J. (2019). Failing grade: 89% of introduction-to-psychology textbooks that define or explain statistical significance do so incorrectly. *Advances in Methods and Practices in Psychological Science, 2*(3), 233–239.

Cohen, J. (1990). Things I have learned (so far). *American Psychologist, 1306*.

Cummins, J. (1978). Bilingualism and the development of metalinguistic awareness. *Journal of Cross-Cultural Psychology, 9*(2), 131–149.

Cummins, J. (1979). Linguistic interdependence and the educational development of bilingual children. *Review of Educational Research, 49*(2), 222–251.

Cummins, J., & Mulcahy, R. (1978). Orientation to language in Ukrainian-English bilingual children. *Child Development, 49*(4), 1239–1242.

Diaz, R. M. (1983). Thought and two languages: The impact of bilingualism on cognitive development. *Review of Research in Education, 10*, 23–54.

Diaz, R. M. (1985). Bilingual cognitive development: Addressing three gaps in current research. *Child Development, 56*, 1376–1388.

Feldman, C., & Shen, M. (1971). Some language-related cognitive advantages of bilingual five-year-olds. *The Journal of Genetic Psychology, 118*(2), 235–244.

Goldberg, B. (1963). On the metalinguistic interpretation of counterfactuals. *The Journal of Philosophy, 60*(11), 291–295.

Goldstein, S., Naglieri, J. A., Princiotta, D., & Otero, T. M. (2014). *Introduction: A history of executive functioning as a theoretical and clinical construct.* In *Handbook of executive functioning* (pp. 3–12). Springer.

Haller, H., & Krauss, S. (2002). Misinterpretations of significance: A problem students share with their teachers. *Methods of Psychological Research, 7*(1), 1–20.

Ianco-Worrall, A. D. (1972). Bilingualism and cognitive development. *Child Development*, 1390–1400.

Lakens, D. (2019, April 9). The practical alternative to the p-value is the correctly used p-value. Retrieved from: https://doi.org/10.31234/osf.io/shm8v

Lykken, D. T. (1968). Statistical significance in psychological research. *Psychological Bulletin, 70*, 151.

MacNamara, J. T. (1966). *Bilingualism and primary education*. The University Press.

Noelting, G. (1980a). The development of proportional reasoning and the ratio concept Part I—Differentiation of stages. *Educational studies in Mathematics, 11*(2), 217–253.

Noelting, G. (1980b). The development of proportional reasoning and the ratio concept Part II—problem-structure at successive stages; problem-solving strategies and the mechanism of adaptive restructuring. *Educational Studies in Mathematics, 11*(3), 331–363.

Nunnally, J. C. (1975). *Introduction to statistics for psychology and education*. Wilson. McGraw-Hill.

Oakes, M. (1986). *Statistical inference: A commentary for the social and behavioural Sciences*. Wiley.

Oren, D. L. (1981). Cognitive advantages of bilingual children related to labeling ability. *The Journal of Educational Research, 74*, 163–169.

Pascual-Leone, J. (1969). *Cognitive development and cognitive style: A general psychological integration* (Unpublished doctoral dissertation, University of Geneva).

Piaget, J. (1929). *The child's conception of the world*. London: Kegan Paul, Trench & Trubner.

Rolstad, K., & MacSwan, J. (2014). The facilitation effect and language thresholds. *Frontiers in Psychology, 5*, 1197.

Smith, E. E. (2001). Cognitive psychology: History. In *International encyclopaedia of the social & behavioral sciences* (pp. 2140–2147). Elsevier.

Wechsler, D. (1974). *Manual for the Wechsler intelligence scale for children-revised*. The Psychological Corporation.

Zelazo, P. D., & Frye, D. (1997). Cognitive complexity and control: A theory of the development of deliberate reasoning and intentional action. In M. Stamenov (Ed.), *Language structure, discourse, and the access to consciousness* (pp. 113–153). John Benjamins.

CHAPTER 6

The Bilingual Advantage

Abstract The term "Bilingual Advantage" had come to represent a general executive functioning advantage observed for bilinguals. While findings originally looked promising, two key publications released in 2015 drew attention to serious methodological problems in the research body and presented evidence that implied that the literature was affected by a publication and confirmation bias, which favoured significant findings in support of a bilingual advantage.

Keywords Bilingualism • Executive functioning • Attention • Replication crisis

6.1 The Bilingual Advantage

The narrative around the advantages and disadvantages of bilingualism started to change around the turn of the century. The Bilingual Problem had become the Bilingual Advantage, and studies increasingly reported findings that suggested the ability to speak a second language greatly benefitted executive functioning skills. In fact, studies with infants suggested that pre-verbal children benefitted from having bilingual parents. In a series of three eye-tracking experiments, Kovács and Mehler (2009) presented seven-month-old children with visual cues and auditory speech cues, which informed them that a "reward" was about to appear. In the first block of each study, the "reward", a picture of a puppet, would appear on either the right or left side of a screen. For the second block, the side on which the reward appeared changed; thus, if the puppet always appeared on the left side of the screen in the first block of trials, it would consistently appear on the right side during the second block of trials. Kovács and Mehler predicted that bilingual and monolingual children would perform equally well in the first block but that bilingual children would find it easier to switch the side to which they paid attention in the second block. Their findings supported this conclusion: the bilingual children adapted better to the change and looked more reliably at the new "correct" side of the screen in the second block.

Bialystok was able to replicate her previous findings that bilingual children performed better on card sorting tasks (e.g. Bialystok & Barac, 2012; Bialystok & Martin, 2004), with other researchers adding to the evidence that bilingual children of all ages performed better on executive functioning tasks than their monolingual peers (e.g. Carlson & Meltzoff, 2008; Nicolay & Poncelet, 2013; Okanda et al., 2010). A growing amount of research pointed towards benefits for adult bilinguals, too. They were found to be less likely to make involuntary eye movements and to have better interference control (Bialystok et al., 2006; Coderre et al., 2013; Poarch & Van Hell, 2012; Salvatierra & Rosselli, 2011; Woumans et al., 2015). On the surface level, "interference control" may sound relatively abstract. What it means is that bilinguals are generally better able to ignore things in their environment that they should ignore, while at the same time paying attention to what they should be paying attention to. This is useful for a number of reasons. Think, for example of driving. If you are able to ignore distracting features in your environment and focus on traffic, this will make you a safer driver. There is evidence that bilingual children deal better with noisy classrooms, as they appear to block out the noise and consequently find it easier to focus on the lesson or the task at hand (Filippi et al., 2015).

The focus of research concerned with the effects of bilingualism on non-linguistic cognitive abilities had truly shifted from intelligence and meta-cognition to executive functioning. While many studies reported an executive functioning advantage for bilinguals, something researchers struggled to agree on was the definition of executive functioning. There is a general agreement that "executive functions" play a crucial role in how we organise, plan, and prioritise our everyday life (Davis et al., 2011;

Meltzer, 2018). However, the details of the underlying processes of executive functioning remain debated (see, e.g., Demetriou et al., 2019; Kopp, 2012; Zelazo & Carlson, 2012). Bialystok's early studies were based on an executive functioning model proposed by Miyake et al. (2000) (later updated by Miyake & Friedman, 2012) which assumed "executive functioning" to consist of three distinct cognitive processes: Attentional Inhibition (i.e. the ability to inhibit interference from distracting information), Attentional Shifting (i.e. the ability to shift the focus of attention as needed), and Updating (i.e. working memory). The tasks that the Bialystok group chose in earlier studies to test executive functioning were based on the model proposed by Miyake et al., and continue to be used in research by the group today (e.g. Craik et al., 2018; Morales et al., 2013). They have the advantage that performance on the tasks, as measured through reaction time and error rates, can be linked directly to Attentional Inhibition, Attentional Shifting, and Updating, which also allows us to look at how these processes interact (e.g. Bialystok et al., 2004). This approach, based on Miyake et al., and the tasks used to study executive functioning, was also widely adopted by other research groups (e.g. Prior & MacWhinney, 2010; Von Bastian et al., 2016). Thus, the exploration of the effects of bilingualism on executive functioning was guided by the Miyake et al. (2000; Miyake & Friedman, 2012) model.

The debate as to whether this model of executive functioning is the most suitable theoretical framework to conceptualise attentional processes and working memory is beyond the scope of the present work. However, the choice of executive functioning model used in earlier studies on the topic still informs the language we use to describe research in this area, and as such we will often see references to Attentional Inhibition, Attentional Shifting, or Updating in the literature, contrary to terminology used in other areas of research concerned with attention or memory.

In the early 2000s, it increasingly appeared that bilingualism almost certainly enhanced executive functioning. An early meta-analysis of research investigating the effect of bilingualism on executive functioning was conducted by Adesope et al. (2010) and concluded that there was a strong body of evidence to suggest bilingualism benefitted executive functioning skills. However, this analysis was limited in the studies it included and was subject to some methodological concerns. More recent meta-analytic work has arrived at different conclusions and raised a number of questions (de Bruin et al., 2015a; Donnelly et al., 2019). These reports raised concerns that the research on bilingualism and executive functioning may be subject to a confirmatory bias (i.e. studies reporting a bilingual advantage are more likely to be published) as well as selective reporting. This posed the question whether bilingualism has any effect on executive functioning at all, or whether previous effects largely appeared credible because of selective publication of studies.

6.2 Confirmatory Bias and Selective Reporting

Surely, if the effects of bilingualism are so powerful that seven-month-old infants exhibit signs of enhanced attentional skills, and studies confirm a bilingual executive functioning advantage across all ages—including for children (e.g. Martin-Rhee & Bialystok, 2008), young adults (e.g. Mor et al., 2015), and healthy older adults (e.g. Schröder & Marian, 2012)—we can be reasonably certain that the effect is reliable. However, this narrative ignores two important aspects of the research it discusses. First, there were flaws and weaknesses within the presented research, for example poorly matched participant groups and a lack of convergent validity across tasks used in individual studies. Second, it ignores research that reports no differences between bilinguals and monolinguals, or that points in the opposite direction, towards a monolingual advantage, resulting in a confirmatory bias.

Selective reporting can lead to the impression of a robust and large bilingual executive functioning advantage. However, it can also be used to create the impression of a monolingual advantage. If someone wanted to highlight the benefits of being monolingual, they would likely mention that monolinguals are less likely to give in to involuntary eye movements and perform better on interference control tasks (e.g. Paap & Greenberg, 2013; Paap & Sawi, 2014; Vu et al., 2010). Being monolingual even appears to balance out differences in white-matter connectivity. When Luk et al. (2011) compared healthy older adult monolinguals and bilinguals in terms of white-matter connectivity, they found that, on average, this was higher in bilinguals, which suggests that they should have performed better than the monolingual group on the administered interference control task, which had been included in a battery of neuropsychological tests. Yet, both groups of participants performed equally well on the task, suggesting that the monolingual executive functioning advantage balanced out the biological disadvantage. However, no one, including Luk et al., has claimed that a monolingual executive functioning advantage exists. Even Paap et al. (e.g. Paap & Greenberg, 2013; Paap & Sawi, 2014), who have reported a monolingual advantage for several of the tasks used in their experiments, are careful to emphasise the limitations of their findings.

There is no doubt that selective reporting can be used to create a powerful narrative, and this is, of course, not limited to research. The aforementioned study by Luk et al. (2011) did not actually conclude that the monolinguals had an advantage, either. Instead, they suggested that, even though a bilingual advantage did not emerge at the behavioural level, bilingualism clearly had a positive effect on white-matter connectivity. An alternative explanation was not considered in the article. It is still debated how differences observed at a biological level should be interpreted in the absence of corresponding differences at the behavioural level (e.g. Bub, 2000; Shulman, 1996; Wilkinson & Halligan, 2004). Some would argue that Luk et al.'s findings could still explain behavioural differences observed between bilinguals and monolinguals in other studies. Others would take the more conservative approach and emphasise that differences at the biological level have little meaning if no differences at the behavioural level are observed. What strengthens the latter approach is that brain imaging studies usually only recruit a small number of participants, which decreases the likelihood that their sample is truly representative for the population we intend to study. The reason for this is time and cost; preparing participants for an EEG or MRI takes considerably more time than welcoming them and asking them to start an experiment that is already set up on a computer in front of them. The equipment and personnel needed to procure, maintain, and run the related technological equipment is also considerably higher, which means time available with these resources tends to be limited. It could be that the small sample in a brain imaging study is generally similar and 'a good match' for the sample of participants in a larger behavioural study, which would mean that the brain imaging findings could potentially be used to inform how we interpret the findings of this larger study. But it would be difficult to ascertain whether this is appropriate without comparing the two participant groups on a wide range of background factors (e.g. age, sex, health history, language profile, socio-economic background), which in reality is rarely, if ever, possible.

Luk et al.'s (2011) research focused on the potential effects of lifelong bilingualism on anatomical brain connectivity. It is thus understandable that an in-depth discussion of behavioural research may have been beyond the scope of their article. However, the interpretation of their findings is particularly surprising as two of the authors of this study also contributed to an article that found the opposite pattern of results. Schweizer et al. (2012) conducted a study with bilinguals and monolinguals who were diagnosed with Alzheimer's disease. The brain atrophy observed in the bilingual participants was far more advanced than in the monolingual group. Nevertheless, monolinguals and bilinguals performed the administered cognitive tests equally well and were around the same age when they were diagnosed with Alzheimer's disease. Schweizer et al. concluded that bilinguals' enhanced executive functioning skills delayed the onset of Alzheimer's symptoms, that is, brain atrophy in bilinguals had to progress further in order for the severity of symptoms to match that of monolinguals with less advanced brain atrophy.

In other words, if monolinguals and bilinguals perform equally well at the behavioural level but bilingual brains appear healthier, this is evidence for a bilingual advantage (Luk et al., 2011). If both groups perform equally well at the behavioural level but monolingual brains are healthier, it is also evidence for a bilingual advantage (Schweizer et al., 2012). This inconsistency in how differences observed at the biological level are interpreted, regardless of whether behavioural results align with them, is, however, not limited to these two studies on the topic and appears to be more widespread. The conclusions of such research (e.g. Abutalebi et al., 2015; Bialystok et al., 2005; Gold et al., 2013; Luk et al., 2010) always interpret the findings as support for a bilingual executive functioning advantage. This is just one example of how the prevalent confirmatory bias in the field has affected the interpretation of research findings. It is crucial to recognise how strong this confirmatory bias was, and still is, in order to appreciate why sub-optimal methodological choices and methods often went unnoticed, leading to more research that, to a greater or lesser degree, supported the idea of a bilingual executive functioning advantage even in the face of weak evidence for it.

Something meta-analyses can only account for up to a point is what result is used to support the conclusion of a bilingual advantage, or indeed, not support it. A common theme in the literature is that there is very little consistency in the pattern of findings. This becomes more apparent as we focus on specific tasks. As an example, the Eriksen Flanker Task (Eriksen & Eriksen, 1974) provides four measures that could be used to assess executive functioning: overall reaction time, overall accuracy, the reaction time interference effect (i.e. reaction time difference between congruent and incongruent trials), and the accuracy interference effect (i.e. difference in number of errors between congruent and incongruent trials).

Calvo and Bialystok (2014) used such a flanker task to test children and found that overall accuracy differed between children, as the bilingual children committed fewer errors overall. However, there were no differences in regard to the other three measures of the flanker task. Calvo and Bialystok viewed these findings as support for a bilingual advantage. In a study by de Abreu et al. (2012) described in more detail in Sect. 6.4.2, monolinguals were slower than bilinguals overall and experienced a larger reaction time interference effect, but no differences emerged between bilinguals and monolinguals in relation to accuracy, which was nevertheless considered support for a bilingual advantage. When it comes to "What measure is the best measure?" the answer obviously depends on the task, but in general there appear to be two camps of researchers: those who view measures that take into account baseline performance, for example interference effects, as preferable (e.g. Blumenfeld & Marian, 2014; Paap & Sawi, 2014), and those who will consider any difference in performance potential evidence of a bilingual advantage. This lack of systematic variation between studies is also discussed by Ross and Melinger (2017), who reviewed these inconsistencies across a number of tasks, and concluded that it is unlikely that a bilingual advantage exists at all.

While previous publications had at times criticised research on the bilingual executive functioning advantage, a meta-analysis conducted by de Bruin et al. (2015a) and a literature review by Paap et al. (2015) elicited a wave of responses and reactions from people who had previously conducted research in this area. The two articles appeared within months of each other and received an unusual amount of coverage, with one article published in *The Atlantic* titled "The Bitter Fight Over the Benefits of Bilingualism" (Yong, 2016). The meta-analysis published by de Bruin et al. (2015a) consisted of two parts. In the first part, they reviewed conference abstracts that were published between 1999 and 2012 and dealt with bilingualism and executive functioning. They then looked into whether the studies presented in the conference abstracts were subsequently published in scientific journals based on whether they supported the idea of a bilingual advantage, or reported mixed results, no differences, or a monolingual advantage. While 68% of the studies that supported a bilingual advantage were subsequently published, only 36% of studies that challenged it to a greater or lesser degree were published, suggesting that the publication process was affected by a confirmatory bias. Of the studies reporting mixed results, 50% were published. However, as shown in the above example, some articles tend to interpret mixed results in such a way that they support a bilingual executive functioning advantage, even if evidence for this is relatively weak. Hence, this does not necessarily provide evidence against a confirmatory bias.

The second part of their meta-analysis aimed to calculate a pooled effect size across all studies, the result of which was a small effect of bilingualism on executive functioning. They also found that the set of studies they looked at was likely to have been affected by a publication bias (i.e. studies that reported significant differences were more likely to be published). In other words, studies that confirmed a bilingual advantage and that reported significant findings were significantly more likely to be published than studies with mixed results or no differences between bilinguals and monolinguals, or those that reported a monolingual advantage. Admittedly, the methods used by de Bruin et al. to estimate publication bias have previously been found to be fairly conservative and easily affected by outliers (Ioannidis & Trikalinos, 2007; Lau et al., 2006; Sterne et al., 2000). However, a second analysis conducted by Lehtonen et al. (2018) used more reliable methods to correct for publication bias and found that if the effects of publication bias are taken into account, the bilingual executive functioning advantage vanishes.

Additionally, in his doctoral dissertation, Donnelly (2016) assessed the effect of moderator variables on studies that investigated the effect of bilingualism on tasks that require interference control or the ability to shift attention (i.e. task-switching tasks). In his meta-analysis of task-switching task studies, he found no evidence of an effect of bilingualism on task-switching ability. However, he did find an effect of bilingualism on interference control, primarily driven by Ellen Bialystok's lab in York, Canada (Ellen, 2015). The studies associated with this research group generally reported a larger difference between monolinguals and bilinguals in terms of overall reaction time, but were comparable to the findings of other research groups in relation to interference control, for which effect sizes ranged from null effects to small effects. Donnelly's findings have since been published (Donnelly et al., 2019), and other recent meta-analytic work also repeatedly concluded that if a bilingual executive functioning advantage exists, then it only occurs within very specific and narrow parameters, which is unlikely to be indicative of a cognitive advantage that would have a noticeable impact on people's lives (Gunnerud et al., 2020; van den Noort et al., 2019; Ware et al., 2020).

There could be a number of explanations for the research group effect, such as differences in the populations that were accessed or differences in research protocols. Bialystok's research group has produced a considerable body of work on this topic, as is not uncommon for a scientist who pioneered a research area, and small differences between research groups and their experimental protocols, such as whether reaction times are trimmed or not, may well have an impact on whether a bilingual executive functioning advantage emerges (Zhou & Krott, 2016). Bialystok et al. (2015) addressed some of these potential explanations in a reply to de Bruin et al.'s (2015a) meta-analysis, such as that data may have been obtained in subpar conditions (e.g. under the time constraints of a dissertation project), potentially making it less likely to find evidence of a bilingual advantage. However, their response to the findings also highlighted deep-rooted differences in the approaches and preferences of scientists when it comes to publishing mixed or non-significant research results. Bialystok et al. go as far as to accuse de Bruin et al. (2015a) of trying to discredit the bilingual advantage:

> There is an understandable preference for journals to publish studies that show some effect over those that show no effect—imagine the state of journals if these studies were published with the same frequency—and this preference may be particularly strong in the case of striking new findings. It seems, therefore, that the real purpose of the de Bruin et al. article is to use publication bias as a means of discrediting evidence for bilingual effects on cognition. (p. 1)

In the light of the history of research that looked at the effects of bilingualism on non-linguistic cognition, it is a legitimate concern to worry about discrediting the bilingual advantage. Language, with all its social, political and economic importance, can be an emotional subject, and if there is no bilingual advantage, people may, rightly or wrongly, worry that there is a bilingual disadvantage. This in turn could lead to a decline in bilingualism or a resurgence of minimising the importance of minoritised languages. For a full response to Bialystok et al. (2015), please see de Bruin et al. (2015b), who responded to the concerns raised in the commentary.

The response by Bialystok et al. (2015) also highlighted an important difference in how researchers view mixed findings. There is a growing trend in psychology to move away from publishing primarily significant and/or novel findings (e.g. Francis, 2012; Kühberger et al., 2014). The rationale behind this is that if all findings on a topic are published, we could avoid or at least minimise the effects of confirmation bias and thus lower the risk of effects appearing larger than they actually are. However, the concerns around how to manage this amount of information are not unwarranted. Historically, significant findings were often viewed as a sign of high-quality research. After all, what are the odds that we find significant differences between conditions or groups if the null hypothesis is true? Traditionally, in psychology, there is less than a 5% chance of this being the case, so if an article reported a significant finding, it was, and still is, often treated at face value, especially if the surrounding research supported the conclusion it presented.

Of course, reporting significant findings alone would not automatically mean a study would be published, but it was viewed as one of the more important markers of good and robust research. Without this rule of thumb, we need a new way to make a balanced decision of what should, and should not, be published. And this becomes increasingly difficult as the number of scientific publications, including in psychology, has exploded since the 1980s (Bornmann & Mutz, 2015; To & Yu, 2020). In other words, more research is being produced, and therefore more research requires peer review and is being published. In itself, this is not a bad thing and holds a great deal of potential, but as our time on any given day is limited, it will become increasingly more difficult to stay up-to-date on the current literature and for reviewers and readers to decide what research meets quality standards—and what should be rejected. This is, partially, why scientists increasingly seem drawn towards Open Science practices, such as data and code sharing, and large-scale collaborative work. These approaches allow for the optimisation of research methods and data analyses, and often mean that the collective knowledge of a much larger team of scientists is available throughout the research process (e.g. Open Science Collaboration, 2012; Prike, 2022).

The literature review by Paap et al. (2015), similar to de Bruin et al. (2015a), also concluded that the research on bilingualism and its effects on executive functioning was subject to a confirmatory bias, but they also raised concerns in regard to confounding variables. They found that background factors that are predictive of executive functioning task performance, such as socio-economic status, were rarely controlled for, potentially creating a bilingual advantage for executive functioning tasks where there is none. Even if researchers claimed that socio-economic status was accounted for as part of the study design or analysis, this is not always the case. For example, Martin-Rhee and Bialystok (2008) claimed that socio-economic status could have no effect on the data, as "all children lived in the same neighbourhood". Yet, we could reasonably expect that even within the same local area there will be substantial differences between households in terms of education, income, and occupation.

The effects of socio-economic status are addressed in more detail in Sect. 6.4.2. However, Paap et al. (2015) also analysed the distribution of findings in favour of a bilingual advantage and those that were not. They found that the distribution of published findings was consistent with a null effect. There are some aspects of research we cannot pick up on by means of quantitative tests, such as the quality of a study (e.g. consistency of testing environments, accuracy of measurement tools), and thus these findings should not be treated as an absolute truth; however, they give reason for concern. Importantly, Paap et al. criticised that executive functioning tasks have low convergent validity, that is, even if bilinguals perform better on one or two executive functioning tasks, they may perform worse or the same as monolinguals on others. While this is concerning in regard to the claim that bilinguals have a universal executive functioning advantage, inconsistencies in the tasks that were used to test executive functioning could be the reason that quantitative measures of the overall effect are weaker or consistent with a null effect.

Between the findings reported in Paap et al.'s (2015) review of the existing literature and de Bruin et al.'s (2015a) findings of a publication and confirmation bias, later to be confirmed by further work, 2015 would become a turning point in how research on bilingualism would be discussed in the literature. Paap et al.'s article in particular elicited a wide range of comments from researchers in the field (e.g. Gade, 2015; Gathercole, 2015; Hartsuiker, 2015; Jared, 2015; Kempe et al., 2015; Morton, 2015; Woumans & Duyck, 2015). One concern that was widely shared was the inappropriate use of statistical methods, such as misinterpreting statistical interactions and relying on covariates to 'control' for interference from confounding variables instead of matching groups of bilinguals and monolinguals on the relevant measures. While there can be legitimate reasons to attempt to control the influence of confounding variables as part of the statistical analysis, the 'gold standard' should be to match participant groups on relevant background factors. However, this is not always possible, and one such area of research is how the effects of bilingualism on executive functioning may affect symptoms of mild cognitive impairment and dementia.

6.3 BILINGUALISM AND DEMENTIA

One major strand of research on bilingualism and executive function has focused on how a potential bilingual advantage may delay the onset of dementia symptoms. There are, generally, two approaches to study these conditions. The first is to enroll a large number of healthy older adults and follow up with them sporadically over a long period of time. Some will develop symptoms of mild cognitive impairment or dementia, while others will remain healthy. It would be impossible to match participants for all relevant background factors ahead of conducting the study, as there is no certain way to predict who will and will not develop dementia—and if there were, we would be faced with considerable ethical considerations. Such ethical considerations apply to the second approach commonly used, for which clinical populations already diagnosed with mild cognitive impairment or dementia are accessed. The ethical concerns around this can impact recruitment number, as participating in research may not be in the person's best interest, and in many cases, researchers will have to rely on information provided by carers or relatives, which is not always accurate. Thus, recruitment can be difficult and often there won't be a sufficient number of participants available to match them on background factors.

The specific causes of different types of dementia remain under investigation but there are some symptoms of dementia that apply almost universally. These include memory loss, or memory problems, at a cognitive level, and brain atrophy at a biological level. Brain atrophy, that is, brain shrinking, is a normal part of aging but progresses faster and to a greater extent in people with dementia. However, atrophy related to different types of dementia tends to be localised, with some brain areas affected more than others. For example, early onset Alzheimer's disease appears to primarily affect the temporoparietal region of the cortex, while atrophy observed for late onset Alzheimer's disease seems to predominantly affect the medial temporal lobe region (Double et al., 1996; Harper et al., 2017). Meanwhile, atrophy observed for frontotemporal dementia primarily affects the temporal lobes and/or frontal lobes, as the name suggests (Rosen et al., 2002; Warren et al., 2013). Thus, while we know that brain atrophy is a biological marker of dementia, where it is localised and at what speed it progresses vary between types of dementia and individuals.

Onset of dementia symptoms has repeatedly been shown to be delayed in bilinguals by four to five years (Alladi et al., 2013; Bialystok et al., 2007;

Craik et al., 2010; de Leon et al., 2020). However, in Anderson et al.'s (2020) recent meta-analysis and review of the topic, they found only a small effect of bilingualism on the age of onset of Alzheimer's disease. They also found that, on balance, bilinguals developed symptoms related to the disease at an older age than monolinguals. This effect remained present even if education and socio-economic status were controlled for.

Notably, however, Anderson et al. (2020) found no evidence to suggest that bilingualism reduces the likelihood of experiencing Alzheimer's disease. While this sounds very promising, one of the issues in establishing whether bilingualism has a preventative effect in regard to Alzheimer's disease specifically, or dementia more generally, is that many of these studies assess 'age of onset' retrospectively, by talking to family members and carers. Mukadam et al. (2017), for example, conducted a meta-analysis similar to that of Anderson et al. but, in contrast to them, split the research into two groups: studies that assessed age of onset retroactively and predictive studies. For studies assessing age of onset retroactively, Mukadam et al. report findings similar to those reported by Anderson et al., that is, the onset of symptoms of Alzheimer's disease was delayed by approximately four to five years. However, for the predictive studies, there was no difference between bilinguals and monolinguals. This strongly suggests that studies that reported a delay of dementia symptoms in bilinguals may have done so as a result of inaccurate age of onset estimates reported by relatives and carers.

What causes these differences in estimates is not clear, however. Without a specific event to mark the occurrence of first symptoms of Alzheimer's disease, the people providing this information may well underestimate or overestimate how long ago they set in. In this case, however, the estimate can be expected to be equally as accurate or inaccurate for bilinguals and monolinguals. There is, of course, a chance that age of onset was over-/underestimated for only one group in a study and that this led to exaggerated results. Nevertheless, only two of the 26 studies identified by Anderson et al. (2020) point towards a comparatively delayed onset of symptoms for monolinguals, while 20 suggest that bilingualism delays the onset of symptoms (the remaining four studies suggest no differences between groups). It is unlikely that the age of onset was underestimated for bilinguals only across all 20 of these studies, which suggests that bilingualism does have some positive effect on cognition in old age, and may delay the onset of dementia symptoms. However, the extent to which it does so remains debated.

6.4 To Match Groups or Not Match Groups

While there are several reasons why we cannot reliably match groups of bilinguals and monolinguals on all confounding variables for research focused on dementia and mild cognitive impairment, for other populations, we should be able to do so. We know that factors such as dialects and culture were linked to how bilingual and monolingual participants performed on intelligence tests. For executive functioning, we see much the same, with immigration, socio-economic status, and type of bilingualism often mentioned as the most important background factors participants should be matched on or that should be controlled for. The following sections will briefly summarise the evidence in regard to the importance of immigration status, socio-economic status, and language profile, before turning to some of the less discussed, potentially confounding, variables.

6.4.1 Immigration

Immigrants, on average, have better health than non-immigrants. They have also been found to have higher cognitive ability than non-immigrants, even before they emigrate from their birth country (Fuller-Thomson et al., 2015). This is sometimes referred to as the 'healthy migrant hypothesis'. Immigrants are also more likely to be bilingual. For example, the Canadian Census in 2016 found that 27.5% of non-immigrants spoke more than one language, compared to 76.4% of Canada's immigrant population (Statistics Canada, Census of Population, 2016). Thus, if by chance a group of bilinguals also includes a large number of immigrants, and is compared to a group of monolinguals that consists mostly of non-immigrants, the bilingual group would be expected to perform better.

That being said, there are some studies that found a bilingual advantage in non-immigrant samples. For example, Bialystok and Viswanathan (2009) compared English-speaking monolingual children in Canada to a mixed group of bilinguals in Canada and a group of Indian bilingual children who spoke English and either Tamil or Telugu. The children were all educated in private schools, with the two groups of Canadian children attending the same school. The implication appears to be that this means that the children were of comparable socio-economic background. Bialystok and Viswanathan acknowledged that the parents of the children in their study may have been immigrants in some cases, but quickly move on to state that, as the children themselves were not immigrants, this should not affect them. However, as immigrants were found to be healthier and to have better cognitive skills prior to moving to a different country, it is likely that at least some of the children would have benefitted from a genetic predisposition towards these characteristics (Fuller-Thomson et al., 2015). Bialystok and Viswanathan did find that the two bilingual groups performed executive functioning tasks much more efficiently than the monolingual group. However, it could be argued that, while some attempts were made to match bilinguals and monolinguals for immigration, the method of matching the children for immigration status has considerable room for improvement.

Additionally, in the first of three studies, Martin-Rhee and Bialystok (2008) matched English-speaking monolingual children and French-English bilingual children for immigration and report results that suggest that the bilingual group performed the administered executive functioning task more efficiently. Moreover, Costa et al. (2008) tested young adult monolinguals and Catalan-Spanish bilinguals, neither group including any immigrants. They found evidence that suggests that the bilinguals in their study resolved conflict between competing attentional demands with more ease than the monolinguals. Thus, there is at least a small number of studies that reported a bilingual advantage for non-immigrant participants.

Other studies find a bilingual advantage for some measures but not others. For example, Kousaie and Phillips (2012) tested young adult and older adult bilinguals and monolinguals without an immigration background. They found that the young bilinguals in their study were overall faster, but in relation to the measure that was indicative of attentional inhibition (i.e. the Stroop effect), there was no difference between bilinguals and monolinguals for either of the two age groups. Gathercole et al. (2010) tested English monolinguals, and three groups of bilinguals. The bilinguals differed based on the language they spoke at home: Welsh, English, or Welsh and English. Gathercole et al. tested both primary school-aged children and teenagers on a Stroop Task, which requires attentional inhibition to be performed successfully. Their participants did not have an immigration background. For teenagers, they found no differences between the four language groups, and for the younger group of children, they found that the monolingual group and the bilinguals who spoke primarily Welsh at home performed worse than the other two groups. As Gathercole et al. acknowledged themselves, if the findings were linked to the primarily Welsh-speaking children being more like monolinguals than bilinguals in their language use, then the children who predominantly spoke English at home should have been subject to the same influences. Yet this was not the case. Their findings cannot support the conclusion that there is a systematically occurring bilingual advantage in non-immigrant populations.

Later on, Gathercole et al. (2014) followed up on these findings by testing all age groups, from three years old to older adults, including a total of 1561 participants. They used the same categorisation of language groups based on home language use and their sample of participants consisted of non-immigrants. The bilinguals they tested were simultaneous and early bilinguals, yet they found no evidence of a bilingual executive functioning advantage. In another large-scale study, Duñabeitia et al. (2014) recruited 504 participants, including Spanish monolinguals and Basque-Spanish bilinguals. Again, neither group had an immigration background and they recruited participants from two age groups, primary school-aged children and teenagers. Duñabeita et al. did not find any evidence for a bilingual advantage, again calling into question whether we can expect to reliably observe a bilingual advantage in non-immigrant populations.

Furthermore, Bialystok et al. (2010) reported mixed findings in regard to bilinguals' executive functioning advantage. They had tested French-speaking monolinguals in France, English-speaking monolinguals in Canada, and French-English bilinguals in Canada. The two non-linguistic executive function tasks they administered were a tapping task and the attentional network task, which requires attentional inhibition to be performed effectively. While the bilinguals performed better on the tapping task, there were no differences between groups for the attentional network task. The authors focused on the significant bilingual advantage on the two linguistic tasks that required attentional control and the tapping tasks, and concluded that their findings are evidence that the bilingual advantage exists regardless of immigration status and culture. What stands out in this article is that the authors administered an extensive, seven-page-long questionnaire to assess children's use of language in different social situations and their proficiency, but instead of asking their parents to provide any indication of their socio-economic status, they dedicated a whole paragraph to stating that it is not necessary to control for this, as there is no evidence that language learning is linked to socio-economic status. This was partly in response to Morton and Harper's (2007) findings, which we will discuss in more detail shortly, who reported data that clearly showed that socio-economic status and not bilingualism affected their participants' performance on executive functioning tasks. Yet, studies that find a bilingual advantage in the absence of an immigration variable do not appear to control for socio-economic status, while the findings of those that did control for it do not support the conclusion of a bilingual advantage.

Of course, when it comes to immigration, we also need to consider potential effects of culture. Culture has been shown to affect executive functioning. Chinese children, for example, consistently perform better on executive functioning tasks than American children of the same age (Lan et al., 2011; Sabbagh et al., 2006). While some articles occasionally state brief observations in regard to participants' cultures, there is rarely a detailed description or discussion of participants' culture available. For example, Carlson and Meltzoff (2008) mentioned that the Hispanic parents of the bilingual children they tested appeared to place greater emphasis on self-control in raising their children, but they did not discuss this observation any further.

A more systematic approach to how culture and bilingualism may interact was provided by Tran et al. (2015). They recruited children from East Asian (Vietnam), Western (Texas), and Western Hispanic (Argentina) cultural backgrounds. In the case of the Western Hispanic children, the authors only recruited monolingual children but the East Asian and Western groups included both bilingual and monolingual children. The children in their sample were around three years old. Tran et al. found that most children in Argentina will learn a second language at school, at which point, however, they would have been older than the other children in the sample. They further reported that the bilingual children within the appropriate age range that they would have considered for inclusion in the study had a higher socio-cconomic status. They excluded these children based on existing literature that strongly implied that socio-economic background does affect executive functioning, and as a consequence there were no bilingual children included in the Western Hispanic group.

The task Tran et al. (2015) used was the attentional network task, which aims to assess the three attentional networks proposed by Posner and Petersen (1990): alerting, orienting, and executive control. When they assessed the performance of their participants for each network, they found an effect of culture for the alertness network and the executive functioning network: Western Hispanic and East Asian children showed signs of an advantage over the children in Texas. However, there was no effect of language group observed for the individual networks, which suggests that at least some findings related to the bilingual executive functioning advantage may be the effect of cultural influences and/or immigration rather than a generalised attentional advantage rooted in bilingualism itself.

Colvin and Allen (1923) investigated the effects of immigrants' 'language handicap', that is, the effect of low proficiency, on performance on intelligence tests. Of course, these tests were, in part, developed to identify highly intelligent immigrants, to ensure that only the most promising individuals who tried to enter, for example the United States, could do so (e.g. Brigham, 1923). In that sense, bilingualism, immigration, and cognitive tests share parts of their history. The idea of culture affecting findings related to bilingualism is also not new. Johnson (1953) proposed that a successful cultural assimilation may lead to better language skills and would thus allow bilinguals to overcome the 'language handicap'. MacLeod (1969) also stated that people's societal experience with their languages may affect their language skills and highlighted that cultural differences can have an effect on these. Historically speaking, it is also interesting to see Bialystok et al.'s (2010) emphasis on bilinguals performing better than their monolingual peers on a tapping task. In 1923, Saer found that the bilinguals he tested struggled much more with the administered tapping task than the monolinguals he tested. As we know, combined with the findings of other administered tasks, he concluded that bilingualism appears to have a negative effect on intelligence. Eighty-seven years later, Bialystok et al. interpreted their findings, along with those from other administered tasks, to support the claim that bilinguals have an executive functioning advantage.

6.4.2 Socio-economic Status

As mentioned in the previous section, there appears to be some reluctance to consider socio-economic background as part of the research design when it comes to studying the bilingual advantage. Definitions of socio-economic status vary but it generally relates to measures of income (e.g. household income), education, and occupation (White, 1982). Different methods have been used in research to assess participants' socio-economic status, although how reliable and accurate these methods are in establishing participants' socio-economic status varies (Rubin et al., 2014). Morton and Harper (2007) were the first to report a study on the bilingual executive functioning advantage in which socio-economic status was explored in more detail. They found that bilingualism did not affect how children performed on the executive functioning task that was given to them but higher socio-economic status was linked to better performance on the task. Following on from Morton and Harper, Carlson and Meltzoff (2008) used a battery of cognitive tasks to compare bilinguals with a low socio-economic status to monolinguals and children enrolled in immersion education with a higher socio-economic status. They combined the tasks they used into a composite score that showed a bilingual advantage. When they analysed the data for each individual task, while accounting for differences in socio-economic background as part of the analysis, they found that the bilinguals performed better than monolinguals on the administered card sorting task, a short-term memory task that also draws on inhibition, and a test of non-verbal intelligence. On the remaining six tasks, the bilingual children did not differ from the other two groups. Carlson and Meltzoff (2008) interpreted these findings as tentative support for a bilingual advantage, while acknowledging the limitations of their work, including the lack of control for cultural influences.

Bialystok published a formal response to Morton and Harper (2007) two years after the publication of their findings, in 2009. Within that response, she highlights that Morton and Harper claimed to replicate two previous studies linked to Bialystok's research group, Martin and Bialystok (2003) and Bialystok et al. (2004), yet these two studies differ in their methodology. She further refers to two articles, Martin-Rhee and Bialystok (2008) and Bialystok (2006), which found that small changes to the Simon Task can alter how it is performed. The Simon Task generally presents people with stimuli that require a response that shares features with the target, usually location, or does not share features with it (Simon & Rudell, 1967). The task exists in many different versions. The one used by Morton and Harper required the children to respond to red and green squares, which appeared either on the left or right side of a computer screen. The children had to respond by pressing a key of the same colour, with the red key located on the left and the green key on the right. This allowed for two testing conditions, congruent and incongruent. In congruent trials, the target appeared on the side on which the correct response button was located, for example the red square appearing on the left. In incongruent trials, the two locations would not match, for example a green square appearing on the left still required children to press the green key that was located to their right. Other task versions ask participants to respond to the words "right" and "left", or present them with arrows pointing in different directions (e.g. Lu & Proctor, 1995; Simon & Rudell, 1967; Watanabe et al., 2015). Performance on the overall task (i.e. across the congruent and incongruent conditions) is thought to be indicative of conflict monitoring abilities, while the difference between the congruent and incongruent conditions is thought to be indicative of interference control (Hilchey & Klein, 2011).

In her response to Morton and Harper (2007), Bialystok highlighted that the children who took part in Morton and Harper's study were about 1.5 years older than in the study they aimed to replicate, and suggested that the children might have outgrown any initial differences that might have been there before. This may of course be true, and it is also true that different versions of the Simon Task may have affected children differently at different ages. The studies she referred to emphasise that small changes to the Simon Task can change the outcome of the task, for example the frequency with which participants had to switch the type of response they made (Bialystok, 2006). A collection of studies published by Martin-Rhee and Bialystok (2008) suggested similar. In the first study reported in this article, the Simon Task had a built-in delay between the stimulus presentation and the opportunity to respond. The delay varied in time, which appears to be the first manipulation she referred to. While bilinguals performed better than monolinguals for some versions of the task, they did not observe a consistent bilingual advantage. The third study reported by Martin-Rhee and Bialystok used arrows instead of squares and found that bilinguals consistently performed better than monolinguals on this task version. However, Bialystok and colleagues did not control for socioeconomic status in any of the studies she refers to, which means that the findings should be interpreted with caution. While this could, theoretically, be due to children's ages and differences in the experimental protocol, if we find a reliable bilingual advantage for the Simon Effect in adults, as, for example, Bialystok et al. (2004) claim, then it is unlikely that children pass through a window of development during which they are not affected by it.

Other points of criticism were raised in the response, and will be briefly acknowledged here. Morton and Harper's (2007) sample size should have been increased; however, the research team around Bialystok has used similar sample sizes. In fact, she used the exact same number of participants in the first study reported in Martin-Rhee and Bialystok (2008), which also deployed a more complex design than Morton and Harper's study and subsequently had even greater need for a larger sample size.

This particular study is an especially interesting comparison as Bialystok (2009) also criticised the unusually long response latencies and the large standard deviation for response times reported by Morton and Harper. However, the findings reported by Study 1 in Martin-Rhee and Bialystok's article shows similarly long response times and standard deviations. Martin-Rhee and Bialystok do not address this as a strength or limitation; it is simply not mentioned. While it is good, and indeed very welcome, to see an active conversation around these topics, especially at a time when samples sizes, and power calculations to estimate them, were not yet widely used, it leaves the question of how warranted this criticism was at the time. One particularly puzzling aspect of Bialystok's response is that she repeatedly cites studies that quite clearly show that socio-economic status and executive functioning are correlated (e.g. Carlson & Meltzoff, 2008; Noble et al., 2005), but at the same time seems to suggest that it is not necessary to control for socio-economic status.

However, sample size does matter. We often describe studies in terms of 'statistical power', that is, the power of a research design to identify the effect we want to study. In other words, if the effect we are interested in really exists, how likely is it that our study design can reveal it? The smaller the effect we want to study, the higher the statistical power we need to find it. Statistical power increases with the number of participants, so if we study a small effect, we need a large sample of participants, while a smaller sample would suffice for a large effect. Let's assume bilinguals really are, on average, a few milliseconds faster or slightly more accurate than monolinguals when completing an executive functioning task. In that case, we would be looking for a small to medium-sized effect, which means that we need a large sample size to find it. A study with a small sample size may, occasionally, 'correctly' find a bilingual advantage but we would not expect to see this advantage consistently across a large number of studies with small sample sizes. Yet this is precisely the pattern we see in meta-analyses on the topic: the larger the sample size, the less likely a study is to report a bilingual advantage (de Bruin & Della Sala, 2019; de Bruin et al., 2015a; Donnelly et al., 2019; Paap et al., 2015). As the statistical power of these smaller studies should have been too low to find a bilingual advantage, this suggests that the effects they report may have occurred by chance, rather than because there was a genuine difference between the bilinguals and monolinguals. We would expect to see this difference for some of them but the large number of small studies revealing a bilingual advantage suggests that there is a publication and confirmation bias present in the literature. Considering this alongside the large studies, which generally do not present evidence of a bilingual advantage, gives reason for concern. I am only aware of one example of researchers intentionally following up on their earlier studies on the topic, which relied on small sample sizes. Filippi and Bright (2022) replicated some of their earlier work with a much larger sample size and found that the effect of a bilingual advantage they had previously observed did not emerge in their later larger study.

The commentary by Bialystok (2009) is important to consider here, as the points raised within it are still at times used to justify not matching participants on socio-economic status. It is also indicative of the approach taken by researchers in this area at the end of the early 2000s, when the focus lay on what details of an experimental protocol would be linked to a bilingual advantage, and which would see it disappear. If the relevance of these details can be explained, this is entirely appropriate. For example, for tasks used to measure the effects of attentional blinks, differences of a few milliseconds can decide whether a participant experiences an attentional blink or not. However, for the executive functioning tasks commonly used to study the effects of bilingualism, such small differences are unlikely to affect the overall pattern of findings. We certainly would not expect a robust effect of bilingualism to disappear because of comparatively minor changes in experimental designs. Yet, this is often what we see, particularly in studies that do not control for background variables.

For a moment, let's assume we can safely accept that children recruited from the same neighbourhood can be considered to have the same socio-economic status. This still does not explain studies that appear to have been conducted across country boundaries, introducing new confounding variables, with no apparent benefit. de Abreu et al. (2012) studied Portuguese monolingual children living in Portugal and Portuguese-Luxembourgish bilinguals living in Luxembourg. The authors did not provide an explanation for why the study was conducted internationally but instead were quick to reassure the reader that the two countries are comparable: "Portugal and Luxembourg are relatively small countries; both are members of the European Union; and there are no apparent within-country disparities in terms of the quality of public school education" (p. 1367). The authors of the study used parental education, household possessions, children's body mass index, and household size to estimate children's socio-economic background. Based on these measures, de Abreu et al. assessed that the bilingual children in Luxembourg had a lower socio-economic status than the monolingual children in Portugal.

The study reported a considerable bilingual advantage and concluded that the bilingual executive functioning advantage balanced out any potential negative effects of socio-economic status on executive functioning. But, of course, this conclusion was based on the assumption that the two countries were indeed comparable.

The article was published in 2012 and it is not clear when data was collected, but if it was within the four years prior to publication, it is relatively safe to assume that data collection was affected by "the Great Recession", which affected both countries quite differently. While Luxembourg went into recession early on, in the second quarter of 2008, the county emerged from the recession only 12 months later, while Portugal spent 51 months in a state of recession between 2007 and 2013 (OECD.Stat, n.d.). Housing prices were higher and competition for housing in Luxembourg was stronger than in Portugal and had been for several years, which calls into question the use of household sizes, which may well be limited by the available accommodation (Stráský, 2020). While statistics for the exact period before 2012 are difficult to come by, in 2015 the gross national income per capita in Portugal was less than half of that of Luxembourg (World Bank, 2016) and Luxembourg's unemployment rate was only half as high as Portugal's (7.75% vs 13.24%; OECD, 2016a). The two countries do indeed perform roughly equally well in terms of how much socio-economic status affects education in schools (OECD, 2015a, 2015b) but the generation of 25–34 year olds, likely to make up some of the parents of the children tested by de Abreu et al., differed in terms of the percentage that attended university: 31.4% in Portugal vs. 52.9% in Luxembourg (OECD, 2016b). Thus, certainly in terms of the economic situation the two countries were not comparable; the "disadvantaged" bilingual children in Luxembourg may very well have benefitted from the national economic security, while the Portuguese monolingual children may have experienced a time of uncertain economic circumstances much more intimately. Whether this is actually the case is difficult to determine, which further poses the question of why the decision was made to test children in different countries, even though we cannot accurately discern whether national differences affected the data.

More recent studies paint a much clearer picture. For example, Vivas et al. (2017) tested Albanian-Greek bilinguals and Greek monolinguals who were matched on socio-economic status and found that both groups performed the executive functioning task given to them equally well. Naeem et al. (2018) tested bilingual and monolingual participants, half of whom had a high socio-economic status and half of whom had a low socio-economic status. They found that bilinguals and monolinguals with high socio-economic status performed equally well. Meanwhile, for participants with low socio-economic status, they found that bilinguals outperformed monolingual participants. However, the problem-solving task administered alongside the executive functioning task revealed a monolingual advantage. Naeem et al. emphasised that these inconsistencies in result patterns run counter to the conclusion that bilingualism leads to advantages in non-linguistic cognitive domains more generally.

Of course, the early revelation that it is not bilingualism that is causing an effect but socio-economic status was also witnessed with intelligence research. Arsenian (1937), for example, delivered fairly strong evidence that socio-economic status, rather than bilingualism, affected people's performance on intelligence tests, but researchers continued to neglect socio-economic status in their research design. For example, Carlson and Henderson (1950) stated that "[t]o control the general socio-economic level and the total cultural complex, only those children who lived in a fairly homogenous, restricted, and older section of Los Angeles were included in the study". Over 70 years on from then, little appears to have changed.

6.4.3 Types of Bilingualism

Earlier on, MacLeod (1969) had suggested that switching between dialects may be similar to switching between languages. More recently, Kirk et al. (2018) set out to study whether this is actually the case for executive functioning. They found that people who regularly switch between dialects of a single language experience a switch cost akin to what bilinguals experience when they switch between languages. By extension, this would suggest that we should also be able to find an executive functioning advantage for anyone who uses more than one dialect. However, research has repeatedly shown that this is not the case, which further calls into question if switching between linguistic systems really leads to a more generalised cognitive advantage (Kirk et al., 2014; Ross & Melinger, 2017; Wu et al., 2016).

As explained in Chap. 1, there are many different types of bilingualism and there is some indication that bilinguals' proficiency may affect whether a bilingual advantage is observed. The more proficient someone becomes in a second language, the more "native like" the way in which they process the language at a cognitive level (Perani et al., 1998), which could suggest that benefits only emerge once bilinguals have learned to successfully manage using two languages.

High-proficient bilinguals appear to perform executive functioning tasks better than low-proficient bilinguals; however, the studies reporting these findings did not control for potentially confounding variables (Bialystok & Craik, 2010; Iluz-Cohen & Armon-Lotem, 2013; Singh & Mishra, 2013; Thomas-Sunesson, Hakuta, & Bialystok, 2018). The pattern of results for differences in bilingual usage, that is, balanced bilinguals and unbalanced bilinguals, has been mixed, and it seems to be generally accepted that proficiency would be expected to have a greater influence on whether a bilingual advantage emerges than bilingual usage (Iluz-Cohen & Armon-Lotem, 2013; Rosselli et al., 2016; Thomas-Sunesson, Hakuta, & Bialystok, 2018).[1] However, there is some evidence that bilingual usage may predict performance on executive functioning tasks to some degree, provided by Weber, Johnson, Riccio, and Liew (2016), who reported a greater bilingual advantage for balanced bilingual children than unbalanced bilingual children. Unfortunately, neither immigration nor socio-economic status was controlled for, although the information required to consider the latter was collected as part of the experiment and reported for the group of participants as a whole. The experiment only compared the two bilingual groups, which means that it is not clear whether monolinguals would in fact have performed worse. Indeed, it could be argued that if significant differences of this kind can be found between children who are fluent in two languages but differ in the frequency and circumstances in which they use them, differences between different groups of bilinguals are just as likely to occur as differences between bilinguals and monolinguals.

While most of the studies that addressed different degrees of proficiency and language use failed to account for confounding variables, it is worth mentioning that Thomas-Sunesson et al. (2018) found that high-proficient bilingual children with a low-socio-economic status performed significantly better on executive functioning tasks than low-proficient children with low socio-economic status. The study did not include children from families with higher socio-economic status or monolinguals,

---

[1] In some of these articles "balanced" and "unbalanced" are used to describe difference in proficiency levels in bilinguals. The definitions applied to analyse the literature are those laid out in Chap. 1.

and it is subsequently not possible to predict how these groups would have performed in comparison to the children with low socio-economic status.

This line of research followed a similar direction as Cummins (1979) when he proposed the threshold hypothesis, according to which bilingual children who manage to become highly proficient in both languages no longer experience a language handicap. While his research aimed to establish whether the link between negative consequences of bilingualism on cognitive development may be limited to only some groups of bilinguals, more recent work did the opposite: What if the inconsistent findings in regard to bilingualism and executive functioning could be explained in terms of only benefitting certain types of bilinguals? What if differences only emerge if we compare 'pure' monolinguals and high-proficient, lifelong, balanced bilinguals to one another? Paap and Sawi (2014) labelled these high-proficient, lifelong, balanced bilinguals the "ideal" bilinguals, but the issue is that, if we look at studies with the ideal bilinguals, including large-scale studies, there is simply no reliable *systematic* evidence that they experience an executive functioning advantage compared to monolinguals (e.g. Antón et al., 2014; Duñabeitia et al., 2014; Gathercole et al., 2014; see also literature review by Ross & Melinger, 2017). In conclusion, while it is possible that the bilingual advantage is limited to certain types of bilinguals, if this is the case, we have yet to identify this type of bilingual.

6.5   Further Considerations: Could We Control for All Potentially Confounding Variables?

While immigration, culture, socio-economic status, and types of bilingualism are important to control for, there are other factors that have been linked to improved executive functioning, for example regular exercise (Verburgh et al., 2014). Intelligence and years spent in education are two other factors linked to enhanced executive functioning: the more intelligent and educated we are, the better our executive functioning (e.g. Lan et al., 2011; Plumet, Gil, & Gaonac'h, 2005). Musical training, either as a vocalist or instrumentalist, has also been reported to convey executive functioning benefits similar to those reported for bilingualism, although it is worth noting that many of these studies also did not control for socio-economic status, which could have affected their outcome (Bialystok & DePape, 2009; Holochwost et al., 2017; Moreno et al., 2011). In many countries around the world musical education, or indeed education more generally, is not available for free to the wider public, which means the ability to pay for musical training, and thus socio-economic status, may well affect the results of these studies. The comparison between musical training and bilingualism was not made for the first time: in the 1970s, Bain (1978) investigated whether musical training and bilingualism benefit mental flexibility, which he found to be the case for both.

This means that if we really wanted to control all possibly confounding variables, we would have to control for socio-economic status, immigration, culture, musical training, exercise, intelligence, and years of education, all while carefully assessing the language profile of participants. These criteria would have to be assessed in addition to ensuring that the sample of participants is sufficiently homogeneous and clearly defined. If we want to compare findings across studies, we need to provide a clear linguistic and non-linguistic profile of participants. If we are not relying on self-reports for proficiency, participants will have to undergo a test to assess linguistic ability in their second language. They would be expected to fill out a profile similar to the LSBQ developed by Anderson et al. (2018), which covers seven pages but would still require an additional questionnaire to assess any experience with sign language. Non-linguistic intelligence tests can take anywhere from 20 minutes to over an hour (e.g. Raven's nine-item version, Bilker et al., 2012; full WAIS–IV, Wechsler, n.d.). Finally, after all of this data is collected we would need to collect the socio-demographic data (age, sex, immigration, culture, musical training). Most executive functioning tasks require much less than 10 minutes to complete, but if we need to record all this information, a quick and convenient task to study executive functioning could quickly turn into an experiment that, at a minimum, would be expected to take an hour. If all of these variables are to be entered into the analysis, the sample size needs to be increased to accommodate the more complex design. If, as an alternative, participants were to be matched on all of these measures, sample size would still need to be increased to at least have a chance at being able to match participants on all of these scales.

There are no studies that have controlled for all of the factors, and it is perhaps not difficult to see why. Some, however, got closer than others, and these are generally linked to an absence of a bilingual advantage. de Bruin et al. (2015), for example, tested older adults and matched them for lifestyle, education, socio-economic status, immigration status, and intelligence when they tested bilinguals and monolinguals, using a Simon Task.

They found that the bilinguals and monolinguals they tested performed the task equally well. Nichols et al. (2020) tested more than 11,000 participants, from 18 to 87 years, on a battery of 12 executive functioning tasks. From these, they were able to match 372 bilinguals and 372 monolinguals for age, education, socio-economic status, gender, and handedness. They also compared the remaining 5994 monolinguals and 5047 bilinguals in an unmatched analysis. For the matched sample, no significant effect of bilingualism was observed for any of three measures. There was an effect of age, which is not unexpected given the age range of the original sample. Crucially, Nichols et al. found that, in the unmatched group of participants, bilinguals had, on average, higher socio-economic status.

This means that the claim that socio-economic status has never been linked to a greater likelihood of being bilingual, and thus does not need to be considered as part of the experimental design, cannot be upheld. The data analysed by Nichols et al. was subject to a self-selection bias, as they acknowledge themselves, and collected online, which means participants' testing environment will have varied widely. While some may argue that this increased ecological validity, others will be concerned about the lack of consistency across participants, and both views are understandable. And while Nichols et al. (2020) included more than 10,000 participants, bilingualism status was self-reported and assessed based on only one question, which could lead to inconsistencies in how participants were categorised. Equally, as only participants from English-speaking countries were included, the conclusions of this study may be limited to this geographic region. However, Nichols et al.'s findings, and indeed those reported by Naeem et al. (2018), Vivas et al. (2017), Morton and Harper (2007), and de Bruin et al. (2015), do not suggest that we should expect to see a difference to emerge between bilinguals and monolinguals.

References

Abutalebi, J., Canini, M., Della Rosa, P. A., Green, D. W., & Weekes, B. S. (2015). The neuroprotective effects of bilingualism upon the inferior parietal lobule: A structural neuroimaging study in aging Chinese bilinguals. *Journal of Neurolinguistics, 33*, 3–13.

Adesope, O. O., Lavin, T., Thompson, T., & Ungerleider, C. (2010). A systematic review and meta-analysis of the cognitive correlates of bilingualism. *Review of Educational Research, 80*(2), 207–245.

Alladi, S., Bak, T. H., Duggirala, V., Surampudi, B., Shailaja, M., Shukla, A. K., et al. (2013). Bilingualism delays age at onset of dementia, independent of education and immigration status. *Neurology, 81*(22), 1938–1944.

Anderson, J. A., Hawrylewicz, K., & Grundy, J. G. (2020). Does bilingualism protect against dementia? A meta-analysis. *Psychonomic Bulletin & Review, 27,* 952–965.

Anderson, J. A., Mak, L., Chahi, A. K., & Bialystok, E. (2018). The language and social background questionnaire: Assessing degree of bilingualism in a diverse population. *Behavior Research Methods, 50*(1), 250–263.

Antón, E., Duñabeitia, J. A., Estévez, A., Hernández, J. A., Castillo, A., Fuentes, L. J., et al. (2014). Is there a bilingual advantage in the ANT task? Evidence from children. *Frontiers in Psychology, 5,* 398.

Arsenian, S. (1937). *Bilingualism and mental development.* College Press.

Bain, B. (1978). The cognitive flexibility claim in the bilingual and music education research traditions. *Journal of Research in Music Education, 26*(2), 76–81.

Bialystok, E. (2006). Effect of bilingualism and computer video game experience on the Simon task. *Canadian Journal of Experimental Psychology, 60*(1), 68.

Bialystok, E. (2009). Claiming evidence from non-evidence: A reply to Morton and Harper. *Developmental Science, 12*(4), 499–503.

Bialystok, E., & Barac, R. (2012). Emerging bilingualism: Dissociating advantages for metalinguistic awareness and executive control. *Cognition, 122*(1), 67–73.

Bialystok, E., Barac, R., Blaye, A., & Poulin-Dubois, D. (2010). Word mapping and executive functioning in young monolingual and bilingual children. *Journal of Cognition and Development: Official journal of the Cognitive Development Society, 11*(4), 485–508.

Bialystok, E., & Craik, F. I. (2010). Cognitive and linguistic processing in the bilingual mind. *Current Directions in Psychological Science, 19*(1), 19–23.

Bialystok, E., Craik, F. I., & Freedman, M. (2007). Bilingualism as a protection against the onset of symptoms of dementia. *Neuropsychologia, 45*(2), 459–464.

Bialystok, E., Craik, F. I., Grady, C., Chau, W., Ishii, R., Gunji, A., & Pantev, C. (2005). Effect of bilingualism on cognitive control in the Simon task: Evidence from MEG. *NeuroImage, 24*(1), 40–49.

Bialystok, E., Craik, F. I., Klein, R., & Viswanathan, M. (2004). Bilingualism, aging, and cognitive control: Evidence from the Simon task. *Psychology and Aging, 19*(2), 290.

Bialystok, E., Craik, F. I., & Ryan, J. (2006). Executive control in a modified antisaccade task: Effects of aging and bilingualism. *Journal of Experimental Psychology: Learning, Memory, and Cognition, 32*(6), 1341.

Bialystok, E., & DePape, A. M. (2009). Musical expertise, bilingualism, and executive functioning. *Journal of Experimental Psychology: Human Perception and Performance, 35*(2), 565.

Bialystok, E., Kroll, J. F., Green, D. W., MacWhinney, B., & Craik, F. I. (2015). Publication bias and the validity of evidence: What's the connection? *Psychological Science, 26*(6), 944–946.

Bialystok, E., & Martin, M. M. (2004). Attention and inhibition in bilingual children: Evidence from the dimensional change card sort task. *Developmental Science, 7*(3), 325–339.

Bialystok, E., & Viswanathan, M. (2009). Components of executive control with advantages for bilingual children in two cultures. *Cognition, 112*(3), 494–500.

Bilker, W. B., Hansen, J. A., Brensinger, C. M., Richard, J., Gur, R. E., & Gur, R. C. (2012). Development of abbreviated nine-item forms of the Raven's standard progressive matrices test. *Assessment, 19*(3), 354–369.

Blumenfeld, H. K., & Marian, V. (2014). Cognitive control in bilinguals: Advantages in Stimulus–Stimulus inhibition. *Bilingualism: Language and Cognition, 17*(3), 610–629.

Bornmann, L., & Mutz, R. (2015). Growth rates of modern science: A bibliometric analysis based on the number of publications and cited references. *Journal of the Association for Information Science and Technology, 66*(11), 2215–2222.

Brigham, C. C. (1923). *A study of American intelligence*. Princeton University Press.

Bub, D. N. (2000). Methodological issues confronting PET and fMRI studies of cognitive function. *Cognitive Neuropsychology, 17*, 467–484.

Calvo, A., & Bialystok, E. (2014). Independent effects of bilingualism and socioeconomic status on language ability and executive functioning. *Cognition, 130*(3), 278–288.

Carlson, H. B., & Henderson, N. (1950). The intelligence of American children of Mexican parentage. *The Journal of Abnormal and Social Psychology, 45*(3), 544–551.

Carlson, S. M., & Meltzoff, A. N. (2008). Bilingual experience and executive functioning in young children. *Developmental Science, 11*(2), 282–298.

Coderre, E. L., Van Heuven, W. J., & Conklin, K. (2013). The timing and magnitude of Stroop interference and facilitation in monolinguals and bilinguals. *Bilingualism: Language and Cognition, 16*(2), 420–441.

Colvin, S. S., & Allen, R. D. (1923). Mental tests and linguistic ability. *Journal of Educational Psychology, 14*(1), 1–20.

Costa, A., Hernández, M., & Sebastián-Gallés, N. (2008). Bilingualism aids conflict resolution: Evidence from the ANT task. *Cognition, 106*(1), 59–86.

Craik, F. I., Bialystok, E., & Freedman, M. (2010). Delaying the onset of Alzheimer disease: Bilingualism as a form of cognitive reserve. *Neurology, 75*(19), 1726–1729.

Craik, F. I., Eftekhari, E., Bialystok, E., & Anderson, N. D. (2018). Individual differences in executive functions and retrieval efficacy in older adults. *Psychology and Aging, 33*(8), 1105.

Cummins, J. (1979). Linguistic interdependence and the educational development of bilingual children. *Review of Educational Research, 49*(2), 222–251.
Davis, C. L., Tomporowski, P. D., McDowell, J. E., Austin, B. P., Miller, P. H., Yanasak, N. E., et al. (2011). Exercise improves executive function and achievement and alters brain activation in overweight children: A randomized, controlled trial. *Health Psychology, 30*(1), 91.
de Abreu, P. M. E., Cruz-Santos, A., Tourinho, C. J., Martin, R., & Bialystok, E. (2012). Bilingualism enriches the poor: Enhanced cognitive control in low-income minority children. *Psychological Science, 23*(11), 1364–1371.
de Bruin, A., Bak, T. H., & Della Sala, S. (2015). Examining the effects of active versus inactive bilingualism on executive control in a carefully matched non-immigrant sample. *Journal of Memory and Language, 85*, 15–26.
de Bruin, A., & Della Sala, S. (2019). The bilingual advantage debate: Publication biases and the decline effect. In J. W. Schwieter & M. Paradis (Eds.), *The handbook of the neuroscience of multilingualism* (pp. 736–753). Wiley & Sons, Incorporated.
de Bruin, A., Treccani, B., & Della Sala, S. (2015a). Cognitive advantage in bilingualism: An example of publication bias? *Psychological Science, 26*(1), 99–107.
de Bruin, A., Treccani, B., & Della Sala, S. (2015b). The connection is in the data: We should consider them all. *Psychological Science, 26*(6), 947–949.
de Leon, J., Grasso, S. M., Welch, A., Miller, Z., Shwe, W., Rabinovici, G. D., et al. (2020). Effects of bilingualism on age at onset in two clinical Alzheimer's disease variants. *Alzheimer's & Dementia, 16*(12), 1704–1713.
Demetriou, E. A., DeMayo, M. M., & Guastella, A. J. (2019). Executive function in autism spectrum disorder: History, theoretical models, empirical findings, and potential as an endophenotype. *Frontiers in Psychiatry, 10*, 753.
Donnelly, S. (2016). *Re-examining the bilingual advantage on interference-control and task-switching tasks: A meta-analysis.* CUNY Academic Works. Retrieved from https://academicworks.cuny.edu/gc_etds/762
Donnelly, S., Brooks, P. J., & Homer, B. D. (2019). Is there a bilingual advantage on interference-control tasks? A multiverse meta-analysis of global reaction time and interference cost. *Psychonomic Bulletin & Review, 26*(4), 1122–1147.
Double, K. L., Halliday, G. M., Krill, J. J., Harasty, J. A., Cullen, K., Brooks, W. S., et al. (1996). Topography of brain atrophy during normal aging and Alzheimer's disease. *Neurobiology of Aging, 17*(4), 513–521.
Duñabeitia, J. A., Hernández, J. A., Antón, E., Macizo, P., Estévez, A., Fuentes, L. J., & Carreiras, M. (2014). The inhibitory advantage in bilingual children revisited. *Experimental Psychology, 61*(3), 234–251.
Eriksen, B. A., & Eriksen, C. W. (1974). Effects of noise letters upon the identification of a target letter in a nonsearch task. *Perception & Psychophysics, 16*(1), 143–149.

Filippi, R., & Bright, P. (2022). A cross-sectional developmental approach to bilingualism: Exploring neurocognitive effects across the lifespan. *Ampersand, 10,* 100097.

Filippi, R., Morris, J., Richardson, F. M., Bright, P., Thomas, M. S., Karmiloff-Smith, A., & Marian, V. (2015). Bilingual children show an advantage in controlling verbal interference during spoken language comprehension. *Bilingualism: Language and Cognition, 18*(3), 490.

Francis, G. (2012). Publication bias and the failure of replication in experimental psychology. *Psychonomic Bulletin & Review, 19*(6), 975–991.

Fuller-Thomson, E., Brennenstuhl, S., Cooper, R., & Kuh, D. (2015). An investigation of the healthy migrant hypothesis: Pre-emigration characteristics of those in the British 1946 birth cohort study. *Canadian Journal of Public Health, 106*(8), e502–e508.

Gade, M. (2015). On tasks and cognitive constructs for the bilingual (non-) advantage. *Cortex, 73,* 347–348.

Gathercole, V. C. M. (2015). Are we at a socio-political and scientific crisis? *Cortex, 73,* 345–346.

Gathercole, V. C. M., Thomas, E. M., Jones, L., Guasch, N. V., Young, N., & Hughes, E. K. (2010). Cognitive effects of bilingualism: Digging deeper for the contributions of language dominance, linguistic knowledge, socio-economic status and cognitive abilities. *International Journal of Bilingual Education and Bilingualism, 13*(5), 617–664.

Gathercole, V. C. M., Thomas, E. M., Kennedy, I., Prys, C., Young, N., Viñas-Guasch, N., et al. (2014). Does language dominance affect cognitive performance in bilinguals? Lifespan evidence from preschoolers through older adults on card sorting, Simon, and metalinguistic tasks. *Frontiers in Psychology, 5,* 11.

Gold, B. T., Kim, C., Johnson, N. F., Kryscio, R. J., & Smith, C. D. (2013). Lifelong bilingualism maintains neural efficiency for cognitive control in aging. *Journal of Neuroscience, 33*(2), 387–396.

Gunnerud, H. L., Ten Braak, D., Reikerås, E. K. L., Donolato, E., & Melby-Lervåg, M. (2020). Is bilingualism related to a cognitive advantage in children? A systematic review and meta-analysis. *Psychological Bulletin, 146*(12), 1059–1083.

Harper, L., Bouwman, F., Burton, E. J., Barkhof, F., Scheltens, P., O'Brien, J. T., et al. (2017). Patterns of atrophy in pathologically confirmed dementias: A voxelwise analysis. *Journal of Neurology, Neurosurgery & Psychiatry, 88*(11), 908–916.

Hartsuiker, R. J. (2015). Why it is pointless to ask under which specific circumstances the bilingual advantage occurs. *Cortex, 73,* 336–337.

Hilchey, M. D., & Klein, R. M. (2011). Are there bilingual advantages on nonlinguistic interference tasks? Implications for the plasticity of executive control processes. *Psychonomic Bulletin & Review, 18*(4), 625–658.

Holochwost, S. J., Propper, C. B., Wolf, D. P., Willoughby, M. T., Fisher, K. R., Kolacz, J., et al. (2017). Music education, academic achievement, and executive functions. *Psychology of Aesthetics, Creativity, and the Arts, 11*(2), 147.

Iluz-Cohen, P., & Armon-Lotem, S. (2013). Language proficiency and executive control in bilingual children. *Bilingualism: Language and Cognition, 16*(4), 884.

Ioannidis, J. P. A., & Trikalinos, T. A. (2007). The appropriateness of asymmetry tests for publication bias in meta-analyses: A large survey. *CMAJ: Canadian Medical Association Journal, 176*(8), 1091–1096.

Jared, D. (2015). What is the theory? *Cortex, 73*, 361–363.

Johnson, G. B. (1953). Bilingualism as measured by a reaction-time technique and the relationship between a language and a non-language intelligence quotient. *The Pedagogical Seminary and Journal of Genetic Psychology, 82*(1), 3–9.

Kempe, V., Kirk, N. W., & Brooks, P. J. (2015). Revisiting theoretical and causal explanations for the bilingual advantage in executive functioning. *Cortex, 73*, 342–344.

Kirk, N. W., Fiala, L., Scott-Brown, K. C., & Kempe, V. (2014). No evidence for reduced Simon cost in elderly bilinguals and bidialectals. *Journal of Cognitive Psychology, 26*(6), 640–648.

Kirk, N. W., Kempe, V., Scott-Brown, K. C., Philipp, A., & Declerck, M. (2018). Can monolinguals be like bilinguals? Evidence from dialect switching. *Cognition, 170*, 164–178.

Kopp, B. (2012). A simple hypothesis of executive function. *Frontiers in Human Neuroscience, 6*, 159.

Kousaie, S., & Phillips, N. A. (2012). Ageing and bilingualism: Absence of a "bilingual advantage" in Stroop interference in a nonimmigrant sample. *Quarterly Journal of Experimental Psychology, 65*(2), 356–369.

Kovács, Á. M., & Mehler, J. (2009). Cognitive gains in 7-month-old bilingual infants. *Proceedings of the National Academy of Sciences of the United States of America, 106*(16), 6556–6560.

Kühberger, A., Fritz, A., & Scherndl, T. (2014). Publication bias in psychology: A diagnosis based on the correlation between effect size and sample size. *PLoS One, 9*(9), e105825.

Lan, X., Legare, C. H., Ponitz, C. C., Li, S., & Morrison, F. J. (2011). Investigating the links between the subcomponents of executive function and academic achievement: A cross-cultural analysis of Chinese and American preschoolers. *Journal of Experimental Child Psychology, 108*(3), 677–692.

Lau, J., Ioannidis, J., Terrin, N., Schmid, C. H., & Olkin, I. (2006). The case of the misleading funnel plot. *British Medical Journal, 333*, 597–600.

Lehtonen, M., Soveri, A., Laine, A., Järvenpää, J., de Bruin, A., & Antfolk, J. (2018). Is bilingualism associated with enhanced executive functioning in adults? A meta-analytic review. *Psychological Bulletin, 144*(4), 394.

Lu, C. H., & Proctor, R. W. (1995). The influence of irrelevant location information on performance: A review of the Simon and spatial Stroop effects. *Psychonomic Bulletin & Review, 2*(2), 174–207.

Luk, G., Anderson, J. A., Craik, F. I., Grady, C., & Bialystok, E. (2010). Distinct neural correlates for two types of inhibition in bilinguals: Response inhibition versus interference suppression. *Brain and Cognition, 74*(3), 347–357.

Luk, G., Bialystok, E., Craik, F. I., & Grady, C. L. (2011). Lifelong bilingualism maintains white matter integrity in older adults. *Journal of Neuroscience, 31*(46), 16808–16813.

MacLeod, F. (1969). *An experimental investigation into some problems of bilingualism* (Doctoral dissertation). Retrieved from http://digitool.abdn.ac.uk/R?func=search-advanced-go&find_code1=WSN&request1=AAIU602195

Martin, M. M., & Bialystok, E. (2003). The development of two kinds of inhibition in monolingual and bilingual children: Simon vs. Stroop. In *Poster presented at the meeting of the Cognitive Development Society, Park City, Utah.*

Martin-Rhee, M. M., & Bialystok, E. (2008). The development of two types of inhibitory control in monolingual and bilingual children. *Bilingualism: Language and Cognition, 11*(01), 81–93.

Meltzer, L. (Ed.). (2018). *Executive function in education: From theory to practice.* Guilford Publications.

Miyake, A., & Friedman, N. P. (2012). The nature and organization of individual differences in executive functions: Four general conclusions. *Current Directions in Psychological Science, 21*(1), 8–14.

Miyake, A., Friedman, N. P., Emerson, M. J., Witzki, A. H., Howerter, A., & Wager, T. D. (2000). The unity and diversity of executive functions and their contributions to complex "frontal lobe" tasks: A latent variable analysis. *Cognitive Psychology, 41*(1), 49–100.

Mor, B., Yitzhaki-Amsalem, S., & Prior, A. (2015). The joint effect of bilingualism and ADHD on executive functions. *Journal of Attention Disorders, 19*(6), 527–541.

Morales, J., Calvo, A., & Bialystok, E. (2013). Working memory development in monolingual and bilingual children. *Journal of Experimental Child Psychology, 114*(2), 187–202.

Moreno, S., Bialystok, E., Barac, R., Schellenberg, E. G., Cepeda, N. J., & Chau, T. (2011). Short-term music training enhances verbal intelligence and executive function. *Psychological Science, 22*(11), 1425–1433.

Morton, J. B. (2015). Still waiting for real answers. *Cortex, 73*, 352–353.

Morton, J. B., & Harper, S. N. (2007). What did Simon say? Revisiting the bilingual advantage. *Developmental Science, 10*(6), 719–726.

Mukadam, N., Sommerlad, A., & Livingston, G. (2017). The relationship of bilingualism compared to monolingualism to the risk of cognitive decline or dementia: A systematic review and meta-analysis. *Journal of Alzheimer's Disease, 58*(1), 45–54.

Naeem, K., Filippi, R., Periche-Tomas, E., Papageorgiou, A., & Bright, P. (2018). The importance of socioeconomic status as a modulator of the bilingual advantage in cognitive ability. *Frontiers in Psychology, 9*, 1818.

Nichols, E. S., Wild, C. J., Stojanoski, B., Battista, M. E., & Owen, A. M. (2020). Bilingualism affords no general cognitive advantages: A population study of executive function in 11,000 people. *Psychological Science, 31*(5), 548–567.

Nicolay, A. C., & Poncelet, M. (2013). Cognitive advantage in children enrolled in a second-language immersion elementary school program for three years. *Bilingualism: Language and Cognition, 16*(3), 597.

Noble, K. G., Norman, M. F., & Farah, M. J. (2005). Neurocognitive correlates of socioeconomic status in kindergarten children. *Developmental Science, 8*(1), 74–87.

OECD. (2015a). *Government at a glance 2015.* Country Fact Sheet Luxembourg. Retrieved from https://www.oecd.org/gov/Luxembourg.pdf

OECD. (2015b). *Government at a glance 2015.* Country Fact Sheet Portugal. Retrieved from https://www.oecd.org/gov/Portugal.pdf

OECD. (2016a). *Unemployment rate.* Retrieved August 11, 2016, from https://doi.org/10.1787/997c8750-en

OECD. (2016b). *Population with tertiary education.* Retrieved August 11, 2016, from https://doi.org/10.1787/0b8f90e9-en

OECD.Stat. (n.d.). *Quarterly national accounts: Quarterly growth rates of real GDP, change over previous quarter.* Retrieved from https://stats.oecd.org/Index.aspx?QueryName=350&QueryType=View&Lang=en#

Okanda, M., Moriguchi, Y., & Itakura, S. (2010). Language and cognitive shifting: Evidence from young monolingual and bilingual children. *Psychological Reports, 107*(1), 68–78E.

Open Science Collaboration. (2012). An open, large-scale, collaborative effort to estimate the reproducibility of psychological science. *Perspectives on Psychological Science, 7*(6), 657–660.

Paap, K. R., & Greenberg, Z. I. (2013). There is no coherent evidence for a bilingual advantage in executive processing. *Cognitive Psychology, 66*(2), 232–258.

Paap, K. R., Johnson, H. A., & Sawi, O. (2015). Bilingual advantages in executive functioning either do not exist or are restricted to very specific and undetermined circumstances. *Cortex, 69*, 265–278.

Paap, K. R., & Sawi, O. (2014). Bilingual advantages in executive functioning: Problems in convergent validity, discriminant validity, and the identification of the theoretical constructs. *Frontiers in Psychology, 5*, 962.

Perani, D., Paulesu, E., Galles, N. S., Dupoux, E., Dehaene, S., Bettinardi, V., et al. (1998). The bilingual brain. Proficiency and age of acquisition of the second language. *Brain: A Journal of Neurology, 121*(10), 1841–1852.

Plumet, J., Gil, R., & Gaonac'h, D. (2005). Neuropsychological assessment of executive functions in women: Effects of age and education. *Neuropsychology, 19*(5), 566.

Poarch, G. J., & Van Hell, J. G. (2012). Executive functions and inhibitory control in multilingual children: Evidence from second-language learners, bilinguals, and trilinguals. *Journal of Experimental Child Psychology, 113*(4), 535–551.

Posner, M. I., & Petersen, S. E. (1990). The attention system of the human brain. *Annual Review of Neuroscience, 13*(1), 25–42.

Prike, T. (2022). Open science, replicability, and transparency in modelling. In *Towards Bayesian model-based demography* (pp. 175–183). Springer.

Prior, A., & MacWhinney, B. (2010). A bilingual advantage in task switching. *Bilingualism: Language and Cognition, 13*(2), 253–262.

Rosen, H. J., Gorno-Tempini, M. L., Goldman, W. P., Perry, R. J., Schuff, N., Weiner, M., et al. (2002). Patterns of brain atrophy in frontotemporal dementia and semantic dementia. *Neurology, 58*(2), 198–208.

Ross, J., & Melinger, A. (2017). Bilingual advantage, bidialectal advantage or neither? Comparing performance across three tests of executive function in middle childhood. *Developmental Science, 20*(4), e12405.

Rosselli, M., Ardila, A., Lalwani, L. N., & Vélez-Uribe, I. (2016). The effect of language proficiency on executive functions in balanced and unbalanced Spanish-English bilinguals. *Bilingualism: Language and Cognition, 19*(3), 489.

Rubin, M., Denson, N., Kilpatrick, S., Matthews, K. E., Stehlik, T., & Zyngier, D. (2014). "I am working-class" subjective self-definition as a missing measure of social class and socioeconomic status in higher education research. *Educational Researcher, 43*(4), 196–200.

Sabbagh, M. A., Xu, F., Carlson, S. M., Moses, L. J., & Lee, K. (2006). The development of executive functioning and theory of mind: A comparison of Chinese and US preschoolers. *Psychological Science, 17*(1), 74–81.

Salvatierra, J. L., & Rosselli, M. (2011). The effect of bilingualism and age on inhibitory control. *International Journal of Bilingualism, 15*(1), 26–37.

Schröder, S. R., & Marian, V. (2012). A bilingual advantage for episodic memory in older adults. *Journal of Cognitive Psychology, 24*(5), 591–601.

Schweizer, T. A., Ware, J., Fischer, C. E., Craik, F. I., & Bialystok, E. (2012). Bilingualism as a contributor to cognitive reserve: Evidence from brain atrophy in Alzheimer's disease. *Cortex, 48*(8), 991–996.

Shulman, R. (1996). Interview. *Journal of Cognitive Neuroscience, 8*, 474–480.

Simon, J. R., & Rudell, A. P. (1967). Auditory S-R compatibility: The effect of an irrelevant cue on information processing. *Journal of Applied Psychology, 51*, 300–304.

Singh, N., & Mishra, R. K. (2013). Second language proficiency modulates conflict-monitoring in an oculomotor Stroop task: Evidence from Hindi-English bilinguals. *Frontiers in Psychology, 4*, 322.

Statistics Canada, Census of Population. (2016). *Linguistic integration of immigrants and official language populations in Canada.* Retrieved from https://www12.statcan.gc.ca/census-recensement/2016/as-sa/98-200-x/2016017/98-200-x2016017-eng.cfm

Sterne, J. A. C., Gavaghan, D., & Egger, M. (2000). Publication and related bias in meta-analysis: Power of statistical tests and prevalence in the literature. *Journal of Clinical Epidemiology, 53*(11), 1119–1129.

Stráský, J. (2020). *Policies for a more efficient and inclusive housing market in Luxembourg.* OECD Working Paper. Retrieved from http://www.oecd.org/officialdocuments/publicdisplaydocumentpdf/?cote=ECO/WKP(2020)2&docLanguage=En

Thomas-Sunesson, D., Hakuta, K., & Bialystok, E. (2018). Degree of bilingualism modifies executive control in Hispanic children in the USA. *International Journal of Bilingual Education and Bilingualism, 21*(2), 197–206.

To, W. M., & Yu, B. T. (2020). Rise in higher education researchers and academic publications. *Emerald Open Research, 2*, 3.

Tran, C. D., Arredondo, M. M., & Yoshida, H. (2015). Differential effects of bilingualism and culture on early attention: A longitudinal study in the US, Argentina, and Vietnam. *Frontiers in Psychology, 6*, 795.

Van den Noort, M., Struys, E., Bosch, P., Jaswetz, L., Perriard, B., Yeo, S., Barisch, P., Vermeire, K., Lee, S. H., & Lim, S. (2019). Does the bilingual advantage in cognitive control exist and if so, what are its modulating factors? A systematic review. *Behavioral Sciences, 9*(3), 27.

Verburgh, L., Königs, M., Scherder, E. J., & Oosterlaan, J. (2014). Physical exercise and executive functions in preadolescent children, adolescents and young adults: A meta-analysis. *British Journal of Sports Medicine, 48*(12), 973–979.

Vivas, A. B., Ladas, A. I., Salvari, V., & Chrysochoou, E. (2017). Revisiting the bilingual advantage in attention in low SES Greek-Albanians: Does the level of bilingual experience matter? *Language, Cognition and Neuroscience, 32*(6), 743–756.

Von Bastian, C. C., Souza, A. S., & Gade, M. (2016). No evidence for bilingual cognitive advantages: A test of four hypotheses. *Journal of Experimental Psychology: General, 145*(2), 246.

Vu, K. P. L., Ngo, T. K., Minakata, K., & Proctor, R. W. (2010). Shared spatial representations for physical locations and location words in bilinguals' primary language. *Memory & Cognition, 38*(6), 713–722.

Ware, A. T., Kirkovski, M., & Lum, J. (2020). Meta-analysis reveals a bilingual advantage that is dependent on task and age. *Frontiers in Psychology, 11*, 1458.

Warren, J. D., Rohrer, J. D., & Rossor, M. N. (2013). Frontotemporal dementia. *British Medical Journal, 347*, f4827.

Watanabe, T., Koyama, S., Tanabe, S., & Nojima, I. (2015). Accessory stimulus modulates executive function during stepping task. *Journal of Neurophysiology, 114*(1), 419–426.

Weber, R. C., Johnson, A., Riccio, C. A., & Liew, J. (2016). Balanced bilingualism and executive functioning in children. *Bilingualism: Language and Cognition, 19*(2), 425–431.

Wechsler, D. (n.d.). *Wechsler Adult Intelligence Scale—Fourth Edition (WAIS-IV)*. Retrieved from https://www.statisticssolutions.com/wechsler-adult-intelligence-scale-fourth-edition-wais-iv/

White, K. R. (1982). The relation between socioeconomic status and academic achievement. *Psychological Bulletin, 91*(3), 461.

Wilkinson, D., & Halligan, P. (2004). The relevance of behavioural measures for functional-imaging studies of cognition. *Nature Reviews Neuroscience, 5*(1), 67–73.

World Bank. (2016). *World Development Indicators database*. Retrieved from: http://databank.worldbank.org/data/download/GNIPC.pdf

Woumans, E., Ceuleers, E., Van der Linden, L., Szmalec, A., & Duyck, W. (2015). Verbal and nonverbal cognitive control in bilinguals and interpreters. *Journal of Experimental Psychology: Learning, Memory, and Cognition, 41*(5), 1579.

Woumans, E., & Duyck, W. (2015). The bilingual advantage debate: Moving toward different methods for verifying its existence. *Cortex, 73*, 356–357.

Wu, Y. J., Zhang, H., & Guo, T. (2016). Does speaking two dialects in daily life affect executive functions? An event-related potential study. *PLoS One, 11*(3), e0150492.

Yong, E. (2016). The bitter fight over the benefits of bilingualism. *The Atlantic*. Retrieved from https://www.theatlantic.com/science/archive/2016/02/the-battle-over-bilingualism/462114/

Zelazo, P. D., & Cunningham, W. A. (2007). Executive function: Mechanisms underlying emotion regulation. In J. J. Gross (Ed.), *Handbook of emotion regulation* (pp. 135–158). The Guilford Press.

Zhou, B., & Krott, A. (2016). Data trimming procedure can eliminate bilingual cognitive advantage. *Psychonomic Bulletin & Review, 23*(4), 1221–1230.

CHAPTER 7

Is Bilingualism Good or Bad?

Abstract This chapter recaps the key conclusions in relation to bilingualism, intelligence, and executive functioning. The chapter also briefly considers similarities between fluid intelligence and executive functioning, and theoretical frameworks relevant to the bilingual executive functioning advantage.

Keywords Bilingualism • Executive functioning • Intelligence • Replication crisis

One hundred years have passed since Saer (1923) and Smith (1923) and cautiously suggested that bilingualism may negatively affect intelligence in children. While we can say with some degree of certainty that bilingualism has no effect on non-verbal intelligence (e.g. Baker, 1988; Darcy, 1963; Hakuta, 1986), some researchers seem to think that more evidence is needed to say the same about executive functioning. Since de Bruin et al. (2015) and Paap et al. (2015) published their findings, there has been a growing body of research that suggests that if the bilingual advantage exists at all, it is neither reliably observed nor very robust (e.g. Arizmendi et al., 2018; Dick et al., 2019).

7.1 The Same Old Question?

Some researchers would argue that we are still, in a way, studying the same non-linguistic form of cognition. In Chap. 3, we briefly talked about the history of intelligence tests and the difference between verbal and non-verbal versions. Another type of intelligence we could consider is general fluid intelligence, often shortened to gF. Fluid intelligence refers to our abstract reasoning and problem-solving ability. It has been argued that it is similar to, or the same, as executive functioning, although whether this comparison is appropriate largely depends on what definition of fluid intelligence and executive functioning is used. Blair (2006), for example, argued that the two terms are synonymous and that both processes rely on the same brain areas. However, he also acknowledged that the definitions and terms used to discuss fluid intelligence and executive functioning varied widely.

This could potentially be problematic, as intelligence tests linked to general fluid intelligence (e.g. Raven's Matrices; Bilker et al., 2012) are often used to 'match' participants for non-verbal intelligence or to 'control for' intelligence as part of the analysis in research interested in the bilingual executive functioning advantage (e.g. Filippi et al., 2020; Mor et al., 2015; Sorge et al., 2017). This means that if general fluid intelligence and executive functioning are the same, or very similar, then matching monolingual and bilingual participants on measures of intelligence could potentially erase a bilingual advantage. If this is the case, it would suggest that bilingualism may indeed affect intelligence. It is also often argued that intelligence, socio-economic status, and education interact with one another and are positively correlated. While we know that these three factors indeed interact, we do not yet know for certain how, specifically, they do so (Strenze, 2007). If general fluid intelligence and executive functioning are more or less the same, and we also match participants on education and socio-economic status, it could mean that a bilingual advantage will not emerge where it otherwise would have.

However, Blair's (2006) view that these two cognitive processes are identical is not universally accepted. For example, in response to Blair, Heitz et al. (2006) emphasised that working memory, executive functioning (which they primarily defined as cognitive processes related to attention), and general fluid intelligence are highly correlated but separate constructs. We have already touched on the inconsistencies in how executive functioning is defined in Chap. 6, and the definitions for of fluid intelligence are similarly inconsistent in their details. For example, there is a question of whether working memory should be considered as a cognitive process in its own right or as part of 'executive functioning' as an umbrella term (e.g. Unsworth et al., 2009; van Aken et al., 2016).

Studies that collected data using executive functioning tasks and intelligence tests that rely on general fluid intelligence generally support Heitz et al.'s (2006) suggestion to treat working memory, executive functioning, and general fluid intelligence as three separate concepts. For example, Shahabi et al. (2014) found that short-term memory, but not working memory or executive functioning, predicted general fluid intelligence. Notably, they used a battery of tests to collect data on fluid intelligence, working memory, and executive functioning tasks. In line with the research discussed in Chap. 6, their results showed poor convergent validity across the executive function tasks. Some of the executive functioning and working memory measures they included were significantly correlated with general fluid intelligence, but there was no clear pattern that would suggest that executive functioning, working memory, and general fluid intelligence are identical concepts. They appear to rely *partially* on the same cognitive processes, but the data does not support the approach to treat these three concepts synonymously.

Where research concludes that executive functioning and general fluid intelligence are particularly highly correlated, this is also often driven by working memory being included under the umbrella term of executive functioning, while attentional processes seem less clearly linked to general fluid intelligence (see, e.g., van Aken et al., 2016). Importantly, besides Shahabi et al.'s (2014) findings, we already know that executive functioning tasks have poor convergent validity (Paap et al., 2015). Thus, even if individual studies reported a high correlation for some executive functioning and/or working memory tasks with general fluid intelligence, it would be unlikely that this correlation would be replicated across all executive functioning and working memory tasks. Therefore, these similarities could not be generalised to all forms of executive functioning even if they were present for individual tasks. However, the correlation between general fluid intelligence and executive functioning tasks is not usually addressed in great detail in the bilingualism literature, and we would benefit from a more in-depth discussion of the topic.

7.2 Theoretical Frameworks

The low convergent validity of executive functioning tasks used to study the effects of bilingualism was not news when Paap et al. (2015) highlighted it as a potential problem. When Friedman and Miyake (2004), and later Miyake and Friedman (2012), proposed their model of executive functioning they reported similarly low correlations between the tasks, and thus this has been known for some time. In itself, this is not a problem. Bilinguals may only have an advantage for some aspects of executive functioning but not all; for example, there may be a bilingual advantage for attentional inhibition but not attentional shifting. However, if this is the case, we would expect to see this advantage consistently on the same tasks for the same measures. And as discussed in Chap. 6, this is not what we see in the literature. There appears to be no systematically occuring bilingual advantage we can reliably observe for any of the executive functioning tasks.

In her early studies on the topic, Bialystok and colleagues (Bialystok, 1992; Bialystok & Majumder, 1998) suggested that bilinguals may experience improved executive functioning skills because they constantly need to monitor and manage access to at least two different languages. The assumption was that this process relied on attention, most likely in the shape of attentional inhibition (i.e. inhibiting the non-relevant language) and attentional shifting (i.e. switching languages). Most people use language throughout their day, and hence it was assumed that bilinguals would have more practice with these attentional skills than monolinguals, who would not have to inhibit or switch languages. Different theories on the topic have been suggested over time (e.g. Blumenfeld & Marian, 2014), and while the details vary, they all assume some degree of language competition or conflict between languages in bilinguals.

This leaves us with a question: Is there a theoretical framework that could support the claim that bilinguals perform better on executive functioning tasks? Jared (2015) asked the same in response to the publication of Paap et al.'s (2015) literature review and concluded that there was no strong support for any of the theories proposed to explain how bilinguals manage access to languages, or how this in turn may affect their executive functioning skills. More recently, Blanco-Elorrieta and Caramazza (2021) attempted to provide a theoretical framework that may be suitable to support the notion of a bilingual executive functioning advantage. While they explored the potential roles of language inhibition and organisation, and how these linguistic effects may transfer to non-linguistic domains, they cannot provide conclusive evidence for one specific theoretical framework. Similar calls for a theoretical framework to support the research in this area have been made by other researchers (e.g. de Bruin et al., 2021), but there remains much uncertainty around the theory behind what may or may not cause a bilingual executive functioning advantage to emerge.

Data-driven and exploratory approaches to research can be of great benefit to science. However, after close to three decades of research on bilingualism and executive functioning, we should have sufficient information to make sense of this effect. Perhaps we cannot expect to have explored every detail of how bilingualism affects executive functioning, but it is not unreasonable to think that, by now, we should be able to develop a framework to explain the theory behind the bilingual advantage. To do so, however, we would need to observe systematic variation across studies and a robust bilingual advantage that emerges reliably. Unfortunately, that is not what we see in the literature (e.g. de Bruin, & Della Sala, 2019; Donnelly, Brooks, & Homer, 2019; Lowe et al., 2021). As such, it is very likely that a bilingual executive functioning advantage does not exist, or that it is so limited that it is unlikely to make a difference in people's day-to-day life.

7.3 Looking Ahead

"If English was good enough for Jesus Christ, it's good enough for me" is a quote that is often attributed to a Texan governor, even though the source is not actually known. It does, however, encompass the line of thought that one language should suffice. If large proportions of the world speak English, why should English native speakers bother to learn another language? The only reason why, presumably, would be that they would not be able to communicate with someone if that person didn't speak English. Considering how widely English is spoken in the Western world, why *doesn't* that person speak English? The subtext of this quote is "Is English, the one language I speak, not good enough for you?" Of course, the findings that linked bilingualism to lower intelligence were at points used to justify the closure of bilingual schools and to enforce the use of only one language in social and professional settings (Baker, 1988; Hakuta & Feldman Mostafapour, 1996). These steps were often linked to historic and/or socio-political lines of thought that viewed immigrants and their culture as inferior. Research which suggests that bilingualism does not harm cognition and instead is linked to cognitive benefits may be seen as sufficient to justify the creation of a safety net for some. Even someone who does not believe in the social, professional, and personal benefits of bilingualism will find it difficult to argue with research that shows that bilingualism makes our brains healthier. The problem, however—as we saw in the previous chapter—is that bilingualism does not "make brains healthier".

Within the published responses to Paap et al. (2015), a common theme emerged. Duñabeitia and Carreiras (2015) concluded that it is unlikely that a bilingual executive functioning advantage exists, while Marzecová (2015) suggested that trying to find a bilingual advantage is akin to trying to find the Loch Ness monster. Marzecova also emphasised that significant resources are expended chasing this mythical advantage and that these resources would be better spent on other research areas. This is a concern that Hartsuiker (2015) also shares, suggesting that resources should be redirected towards areas of research that are more likely to yield important information. What they have in common is that most of the reactions to Paap et al.'s article are ready to move on from the bilingual advantage (e.g. Gade, 2015; Gathercole, 2015; Hartsuiker, 2015; Jared, 2015; Kempe et al., 2015; Morton, 2015; Woumans & Duyck, 2015).

Following Saer et al.'s (1924) report of negative effects of bilingualism on intelligence, it took more than 10 years until Arsenian (1937) published his findings that socio-economic status affects performance on intelligence tests. Pintner and Keller (1922) reported that linguistic intelligence tests are not suitable to compare intelligence between bilinguals and monolinguals around the same time that Saer (1923) and Smith (1923) published their findings. Between these two observations, the findings of earlier studies could easily have been explained but researchers continued for decades without controlling for socio-economic status between groups and using linguistic intelligence tests. As we have seen, one of the reasons for this were socio-political beliefs linked to eugenics. In the eyes of some researchers, truly intelligent bilinguals should have been able to adapt to different modalities of intelligence tests and matched their monolingual peers on the verbal versions. It took until the 1960s for the idea that bilingualism causes "mental retardation" to fade into the past (Darcy, 1963). If this history is anything to go by, we may still have several decades of research that investigates the effect of bilingualism on executive functioning to look forward to. I do, however, think we would have let go of the idea that bilingualism benefits executive functioning if historic research had not framed the ability to speak a second language as harmful and if it had not been for the negative consequences of this narrative.

In his report on the Belgian school system, Williams (1915) quoted a director of schools who said that teaching languages is the "plain duty of every statesmen", as it will reduce the divide between different communities. This, of course, has not changed. As mentioned at the beginning of this book, the language we choose can invite someone into the conversation or exclude them. If every politician insisted on only speaking one language and never involved a translator, international politics would be a thing of the past. If the benefits which arise from cultural exchange alone are not sufficient to justify learning another language, however, it might be worth bearing in mind that bilinguals, on average, earn more than monolinguals and are perceived as more physically attractive (Gándara, 2018; Howard, 2016).

References

Arizmendi, G. D., Alt, M., Gray, S., Hogan, T. P., Green, S., & Cowan, N. (2018). Do bilingual children have an executive function advantage? Results from inhibition, shifting, and updating tasks. *Language, Speech, and Hearing Services in Schools, 49*(3), 356–378.

Arsenian, S. (1937). *Bilingualism and mental development.* College Press.

Baker, C. (1988). *Key issues in bilingualism and bilingual education* (Vol. 35). Multilingual Matters.

Bialystok, E. (1992). Selective attention in cognitive processing: The bilingual edge. *Advances in Psychology, 83*, 501–513.

Bialystok, E., & Majumder, S. (1998). The relationship between bilingualism and the development of cognitive processes in problem solving. *Applied Psycholinguistics, 19*, 69–85.

Bilker, W. B., Hansen, J. A., Brensinger, C. M., Richard, J., Gur, R. E., & Gur, R. C. (2012). Development of abbreviated nine-item forms of the Raven's standard progressive matrices test. *Assessment, 19*(3), 354–369.

Blair, C. (2006). How similar are fluid cognition and general intelligence? A developmental neuroscience perspective on fluid cognition as an aspect of human cognitive ability. *Behavioral and Brain Sciences, 29*(2), 109–125.

Blanco-Elorrieta, E., & Caramazza, A. (2021). On the need for theoretically guided approaches to possible bilingual advantages: An evaluation of the potential loci in the language and executive control systems. *Neurobiology of Language, 2*(4), 452–463.

Blumenfeld, H. K., & Marian, V. (2014). Cognitive control in bilinguals: Advantages in stimulus–stimulus inhibition. *Bilingualism: Language and Cognition, 17*(3), 610–629.

Darcy, N. T. (1963). Bilingualism and the measurement of intelligence: Review of a decade of research. *The Journal of Genetic Psychology, 103*(2), 259–282.

de Bruin, A., & Della Sala, S. (2019). The bilingual advantage debate: Publication biases and the decline effect. In Schwieter, J. W., & Paradis, M., *The Handbook of the Neuroscience of Multilingualism* (pp. 736–753). Wiley & Sons, Incorporated.

de Bruin, A., Dick, A. S., & Carreiras, M. (2021). Clear theories are needed to interpret differences: Perspectives on the bilingual advantage debate. *Neurobiology of Language, 2*(4), 433–451.

de Bruin, A., Treccani, B., & Della Sala, S. (2015). Cognitive advantage in bilingualism: An example of publication bias? *Psychological Science, 26*(1), 99–107.

Dick, A. S., Garcia, N. L., Pruden, S. M., Thompson, W. K., Hawes, S. W., Sutherland, M. T., et al. (2019). No evidence for a bilingual executive function advantage in the ABCD study. *Nature Human Behaviour, 3*(7), 692–701.

Donnelly, S., Brooks, P. J., & Homer, B. D. (2019). Is there a bilingual advantage on interference-control tasks? A multiverse meta-analysis of global reaction time and interference cost. *Psychonomic Bulletin & Review, 26*(4), 1122–1147.

Duñabeitia, J. A., & Carreiras, M. (2015). The bilingual advantage: Acta est fabula. *Cortex, 73,* 371–372.

Filippi, R., Ceccolini, A., Periche-Tomas, E., & Bright, P. (2020). Developmental trajectories of metacognitive processing and executive function from childhood to older age. *Quarterly Journal of Experimental Psychology, 73*(11), 1757–1773.

Friedman, N. P., & Miyake, A. (2004). The relations among inhibition and interference control functions: A latent-variable analysis. *Journal of Experimental Psychology: General, 133*(1), 101.

Gade, M. (2015). On tasks and cognitive constructs for the bilingual (non-) advantage. *Cortex, 73,* 347–348.

Gándara, P. (2018). The economic value of bilingualism in the United States. *Bilingual Research Journal, 41*(4), 334–343.

Gathercole, V. C. M. (2015). Are we at a socio-political and scientific crisis? *Cortex, 73,* 345–346.

Hakuta, K. (1986). *Mirror of language. The debate on bilingualism.* Basic Books, Inc..

Hakuta, K., & Feldman Mostafapour, E. (1996). Perspectives from the history and politics of bilingualism and bilingual education in the United States. In I. Parasnis (Ed.), *Culture and language diversity and the deaf experience* (pp. 38–50). Cambridge University Press.

Hartsuiker, R. J. (2015). Why it is pointless to ask under which specific circumstances the bilingual advantage occurs. *Cortex, 73,* 336–337.

Heitz, R. P., Redick, T. S., Hambrick, D. Z., Kane, M. J., Conway, A. R., & Engle, R. W. (2006). Working memory, executive function, and general fluid intelligence are not the same. *Behavioral and Brain Sciences, 29*(2), 135–136.

Howard, L. (2016). Most Americans find this sexy. *Bustle*. Retrieved from https://www.bustle.com/articles/180441-bilinguals-are-more-attractive-say-most-americans-and-heres-why

Jared, D. (2015). What is the theory? *Cortex, 73*, 361–363.

Kempe, V., Kirk, N. W., & Brooks, P. J. (2015). Revisiting theoretical and causal explanations for the bilingual advantage in executive functioning. *Cortex, 73*, 342–344.

Lowe, C. J., Cho, I., Goldsmith, S. F., & Morton, J. B. (2021). The bilingual advantage in children's executive functioning is not related to language status: A meta-analytic review. *Psychological Science, 32*(7), 1115–1146.

Marzecová, A. (2015). Bilingual advantages in executive control–A Loch Ness Monster case or an instance of neural plasticity. *Cortex, 73*, 364–366.

Miyake, A., & Friedman, N. P. (2012). The nature and organization of individual differences in executive functions: Four general conclusions. *Current Directions in Psychological Science, 21*(1), 8–14.

Mor, B., Yitzhaki-Amsalem, S., & Prior, A. (2015). The joint effect of bilingualism and ADHD on executive functions. *Journal of Attention Disorders, 19*(6), 527–541.

Morton, J. B. (2015). Still waiting for real answers. *Cortex, 73*, 352–353.

Paap, K. R., Johnson, H. A., & Sawi, O. (2015). Bilingual advantages in executive functioning either do not exist or are restricted to very specific and undetermined circumstances. *Cortex, 69*, 265–278.

Pintner, R., & Keller, R. (1922). Intelligence tests of foreign children. *Journal of Educational Psychology, 13*(4), 214–222.

Saer, D. J. (1923). The effect of bilingualism on intelligence. *British Journal of Psychology: General Section, 14*(1), 25–38.

Saer, D. J., Smith, F., & Hughes, J. (1924). *The bilingual problem*. University College Wales.

Shahabi, S. R., Abad, F. J., & Colom, R. (2014). Short-term storage is a stable predictor of fluid intelligence whereas working memory capacity and executive function are not: A comprehensive study with Iranian schoolchildren. *Intelligence, 44*, 134–141.

Smith, F. (1923). Bilingualism and mental development. *British Journal of Psychology: General Section, 13*(3), 271–282.

Sorge, G. B., Toplak, M. E., & Bialystok, E. (2017). Interactions between levels of attention ability and levels of bilingualism in children's executive functioning. *Developmental Science, 20*(1), e12408.

Strenze, T. (2007). Intelligence and socioeconomic success: A meta-analytic review of longitudinal research. *Intelligence, 35*(5), 401–426.

Unsworth, N., Miller, J. D., Lakey, C. E., Young, D. L., Meeks, J. T., Campbell, W. K., & Goodie, A. S. (2009). Exploring the relations among executive functions, fluid intelligence, and personality. *Journal of Individual Differences, 30*(4), 194–200.

van Aken, L., Kessels, R. P., Wingbermühle, E., van der Veld, W. M., & Egger, J. I. (2016). Fluid intelligence and executive functioning more alike than different? *Acta Neuropsychiatrica, 28*(1), 31–37.

Williams, J. G. (1915). *Mother-tongue and other-tongue, or a study in bilingual teaching*. Jarvis and Foster.

Woumans, E., & Duyck, W. (2015). The bilingual advantage debate: Moving toward different methods for verifying its existence. *Cortex, 73*, 356–357.

Index[1]

B
Bilingualism
 definition, 6, 13, 20
 scale/spectrum, 49, 57
Bilingual Problem, v, 37–51, 81

C
Confirmation bias, 90, 91, 101
Convergent validity, 84, 91, 121, 122

D
Dialect, 10–14, 20, 28, 30n1, 62, 94, 104

E
Education, 2–5, 10, 20, 27–34, 38–41, 49, 61–65, 91, 93, 98, 102, 103, 106–108, 120

Eugenics, v, 42, 44–46, 124
Executive functioning, v, 2, 6, 62, 70, 77, 81–84, 86–91, 94–99, 101–108, 119–124
Exercise, 106, 107

I
Immersion education, 3, 5, 62, 63, 99
Immigration, 10, 39, 40, 42–45, 47, 94–98, 105–107
Intelligence tests, v, 37–47, 49–51, 56–60, 62, 94, 98, 104, 107, 120, 121, 124
Irish, 3, 63–65

M
Meta-linguistics, 69–77
Musical training, 70, 106, 107

---

[1] Note: Page numbers followed by 'n' refer to notes.

P
Publication bias, 88, 89, 91, 101
P-value, 76

R
Replication crisis, 82, 99, 100, 102, 121

S
Sign language, 13, 14, 20, 107

Socio-economic background, v, 41, 46, 51, 56, 58, 60, 63, 71, 85, 94, 97–99, 102
Statistical significance, 76
Statistics, 76, 103

T
Threshold hypothesis, 70, 71, 106

Ingram Content Group UK Ltd.
Milton Keynes UK
UKHW020047040723
424495UK00005B/103